CIRCUIT ANALYSIS
Theory and Practice 2E

LABORATORY MANUAL

D0074154

Allan H. Robbins
Wilhelm C. Miller

Delmar
Thomson Learning™

Africa • Australia • Canada • Denmark • Japan • Mexico
New Zealand • Philippines • Puerto Rico • Singapore
Spain • United Kingdom • United States

NOTICE TO THE READER

Delmar Staff

Publisher: Alar Elken
Executive Editor: Sandy Clark
Acquisitions Editor: Gregory L. Clayton
Development Editor: Michelle Ruelos Cannistraci
Editorial Assistant: Jennifer Thompson
Executive Marketing Manager: Maura Theriault
Channel Manager: Mona Caron

Executive Production Manager: Mary Ellen Black
Production Manager: Larry Main
Senior Project Editor: Chistopher Chien
Art/Design Director: Nicole Reamer
Marketing Coordinator: Paula Collins
Technology Project Manager: Tom Smith

COPYRIGHT © 2000
Delmar is a division of Thomson Learning. The Thomson Learning logo is a registered trademark used herein under license.

Printed in Canada
1 2 3 4 5 6 7 8 9 10 XXX 05 04 03 02 01 00

For more information contact: Delmar, 3 Columbia Circle, PO Box 15015, Albany, NY 12212-0515; or find us on the World Wide Web at http://www.delmar.com

Asia
Thomson Learning
60 Albert Street, #15-01
Albert Complex
Singapore 189969

Japan
Thomson Learning
Palaceside Building 5F
1-1-1 Hitotsubashi, Chiyoda-ku
Tokyo 100 0003 Japan

Australia/New Zealand:
Nelson/Thomson Learning
102 Dodds Street
South Melbourne, Victoria 3205
Australia

UK/Europe/Middle East
Thomson Learning
Berkshire House
168-173 High Holborn
London WC1V 7AA United Kingdom

Thomas Nelson & Sons LTD
Nelson House
Mayfield Road
Walton-on-Thames
KT 12 5 PL United Kingdon

Latin America
Thomson Learning
Seneca, 53
Colonia Polanco
11560 Mexico D.F. Mexico

Canada
Nelson/Thomson Learning
1120 Birchmount Road
Scarborough, Ontario
Canada M1K 5G4

Spain
Thomson Learning
Calle Magallanes, 25
28015-MADRID
ESPANA

International Headquarters
Thomson Learning
International Division
290 Harbor Drive, 2nd Floor
Stamford, CT 06902-7477

Library of Congress Cataloging-in-Publication Data
ISBN# 0-7668-0627-8

Contents

Preface

This manual provides a set of laboratory exercises that cover the basic concepts of dc and ac circuit theory. While its 37 experiments are more than can be covered in a normal sequence in an introductory course, it provides flexibility and allows instructors to choose labs to suit their programs. The sequence is also flexible. For example, some instructors may wish to move the introductory oscilloscope lab to an earlier spot—immediately following Lab 12, for instance.

Each lab includes a short overview of key ideas, a set of objectives, a list of equipment and parts, instructions on how to carry out the investigation, appropriate tables for summarizing data, and a set of review questions or problems to test the student's comprehension of the material covered. A mini-tutorial and reference guide to equipment and basic measurement techniques is included at the beginning of the manual to provide the student with a readily accessible source of practical information for use during the lab program. A short section on safety in the laboratory follows this guide.

The equipment needed to run these labs is, for the most part, common equipment of the type found at all colleges. Most experiments, for example, can be performed with just a variable dc power supply, digital and analog multimeters, a signal generator, and a two-channel oscilloscope. A few require additional equipment such as an *LRC* meter (or an impedance bridge), a pair of wattmeters, and a three-phase source. A complete list of equipment and components needed is contained in Appendix B.

In addition to the equipment labs, a number of computer-based labs are included for those who teach OrCAD PSpice or Electronics Workbench. While grouped at the end of the manual, they may be inserted where applicable.

While this manual is designed as a companion lab book for the text, *Circuit Analysis: Theory and Practice* by Allan H. Robbins and Wilhelm C. Miller, it may be used with any suitable text.

Allan H. Robbins
Wilhelm C. Miller
August 1999

Acknowledgments

We would like to thank the reviewers. Many of your suggestions have been incorporated in the manual. We would also like to thank the staff at Delmar for their efforts in helping us put this project together. Specifically, we would like to mention Michelle Ruelos Cannistraci, Larry Main, Nicole Reamer, Christopher Chien, and Greg Clayton—a special thanks to all of you.

A Guide to Lab Equipment and Laboratory Measurements

Before we get started, let us take a brief look at lab equipment and basic measurement techniques. If you are unfamiliar with test equipment, you should read this section carefully. For example, before you begin the dc portion of the lab, read the sections on power supplies and voltage, current and resistance measurements. Before you begin the ac experiments, read the sections on the function generator and the oscilloscope.

The Power Supply

Your main source of dc in the laboratory will be a variable voltage, regulated dc power supply. The supply plugs into the ac wall outlet and converts the incoming ac to dc. Such supplies generally include a front panel control for setting the desired voltage and a built-in meter to monitor voltage (and sometimes current). A typical supply is shown in Figure 1. Both single-voltage and multi-voltage units are available. (With multi-voltage units, several sections can usually be operated in series for increased voltage or in parallel for increased current). The typical range of output voltage for a laboratory supply is from zero to 30 volts.

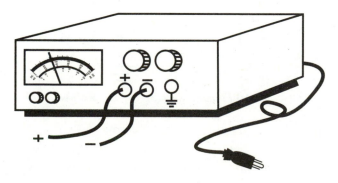

Figure 1   Power supply

Floating Outputs

Most laboratory power supplies have *floating outputs*—that is, outputs where both the positive and negative terminals are isolated from ground. (The source voltage appears between the terminals, and no voltage appears between either terminal and ground.) Such supplies may be operated with one or the other of its terminals jumpered to ground, Figure 2, or both terminals may be left floating. Possible connections are shown in Figure 3.

The Multimeter

One of the most popular test instruments for basic measurements is the *multimeter*. A multimeter combines the functions of a voltmeter, ammeter, and ohmmeter into a single instrument. Both analog and digital units are available. (The digital instrument is known as a digital multimeter or *DMM*, while the analog instrument is known as volt-ohm-milliammeter or *VOM*.) DMMs display results on a numeric readout, while VOMs use a needle pointer and a set of calibrated scales. DMMs are more accurate than VOMs (as we discuss later), but may have a poorer frequency response. (Some DMMs for example can measure only up to about 1 kHz, although others can measure to 100 kHz or higher. For VOMs, the upper frequency limit is typically 100 kHz, although this depends on range. Always check your manual for details.) DMMs generally have higher impedance and result in less loading than VOMs. Figure 4 shows a hand-held DMM and a bench VOM. Note that the type of measurement to be made is selected by means of a function selector switch.

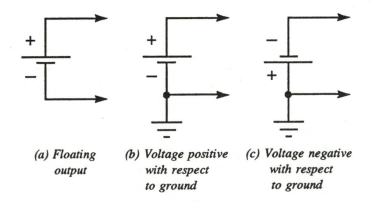

(a) Floating output

(b) Voltage positive with respect to ground

(c) Voltage negative with respect to ground

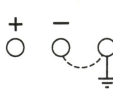

Figure 2 Power supply connections. Either terminal may be jumpered to ground.

Figure 3 Possible power supply connections

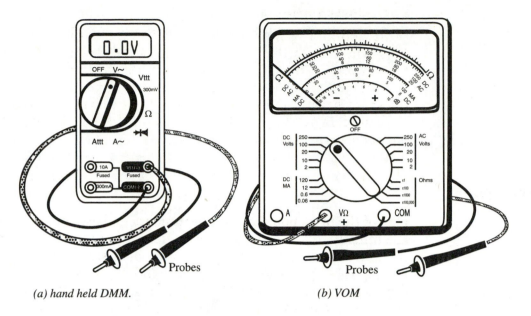

(a) hand held DMM.   (b) VOM

Figure 4   Digital and analog meters. Reproduced with permission from John Fluke Mfg. Co., Inc.

Special Features of the DMM

Two features that many DMMs have are *autoranging* and *autopolarity*. With autoranging, you simply select the desired function (voltage, current, or resistance) then let the DMM automatically determine the correct range. Similarly for polarity—if you connect the instrument for a dc measurement, it automatically determines the polarity (for voltage) or direction (for current) and displays the appropriate sign as part of the measured result. Some DMMs also measure other quantities such as capacitance and frequency.

Manual Range Selection

For VOMs (and non-autoranging DMMs), you must manually select the appropriate range, e.g., 250 V full scale, 120 mA full scale, and so on. (Be sure that you do this before you energize the circuit.) If you have no idea of the magnitude of the quantity to be measured, start at the highest range to avoid possible instrument damage and work your way down to lower ranges until you get the best possible reading.

Terminal Connections

As indicated in Figure 4, a multimeter has two main terminals and several secondary terminals. One main terminal is generally designated *COMMON*, *COM* or minus (–). Designations for the other terminals vary. For example, some meters have a common set of terminals for voltage and resistance (which may be designated *VΩ* or

plus (+) or some similar designation) plus one or more separate terminals for current. Other meters have a combined voltage/resis-
tance/current input labelled $V\Omega A$. Standard lead colors are black and
red; the black lead connects to the (–) or COM terminal while the red
lead connects to the other main terminal, i.e., (+) or $V\Omega$.

Average Reading and True rms Reading Instruments

AC meters are calibrated to read rms values. However, most meters
(called *average responding*) are designed to measure rms for sine
waves only—i.e., they are not capable of determining the rms value
for non-sinusoidal waveforms such as square waves, triangular
waves, superimposed ac and dc, etc. For these, you need a special
DMM called a *true rms* meter. True rms meters are somewhat more
expensive than standard meters.

How To Measure dc Voltage

Voltage is measured as in Figure 5: 1) Ensure that leads are plugged
correctly into the meter sockets—red lead in the *V* socket and the
black lead in the COM socket; 2) Set the function selector to *dc voltage* and select the range if the meter is not autoranging; 3) Connect
the probes across the circuit element whose voltage you wish to measure; 4) Read the voltage. (Although we have shown a hand-held
DMM, the connection is the same for a bench DMM or a VOM.)

*Reproduced with
permission from
John Fluke Mfg. Co., Inc.*

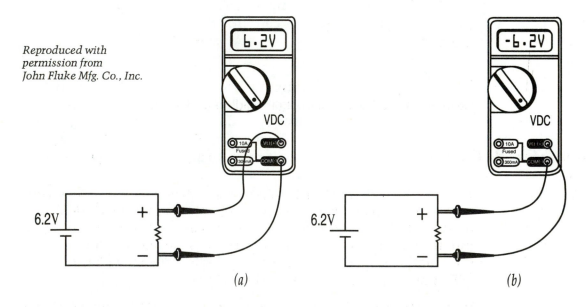

Figure 5   Measuring dc voltage. The sign for (b) is negative.

Points to Note with Respect to Figure 5: If the red (+) lead is connected to the positive side of the circuit and the black (–) lead to the negative side, the meter reading will be positive as in (a). Conversely, if the red lead is connected to the negative side of the circuit and the black lead to the positive side, the meter reading will be negative as in (b) if the DMM has autopolarity. For a VOM, the meter in (b) will try to read down scale. In this case, you will have to reverse the connections or the VOM may be damaged.

How To Measure ac Voltage

The procedure for ac voltage is the same as for dc voltage except that you set the selector dial to *ac voltage*. In this case however, since an ac meter reads magnitude only, it does not matter which way you connect the meter. This is indicated in Figure 6. As you can see, both connections show the same result. This is true whether you are using a digital or an analog instrument.

Reproduced with permission from John Fluke Mfg. Co., Inc.

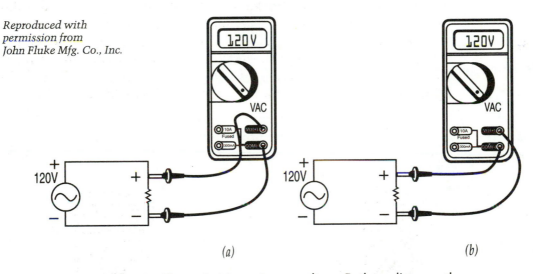

(a)   (b)

Figure 6 Measuring ac voltage. Both readings are the same.

How To Measure dc Current

To measure dc current: 1) Turn power off, open the circuit, then connect so that the current you wish to measure passes through the meter as indicated in Figure 7. Ensure that the test lead is plugged into the current jack if the meter has a separate current input; 2) Set the function selector to dc current and select the range if necessary; 3) Energize the circuit and read the current.

Points to Note: If the current enters at the A (or VΩA) terminal and exits at COM, the reading will be positive as in (a). If the leads are

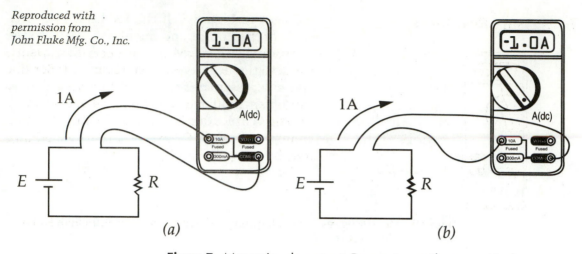

Figure 7   Measuring dc current. Be sure to use the current jack.

reversed as in (b), the reading will be negative. For a VOM, the meter in (b) will try to read down scale. In this case, reverse the connections.

How To Measure ac Current

Note: Not all multimeters measure ac current. For those that do, proceed as for the dc case, except set the dial to ac current instead of dc. (Since an ac meter reads magnitude only, you can connect as in Figure 8(a) or (b). This is true whether you are using a digital or an analog instrument.)

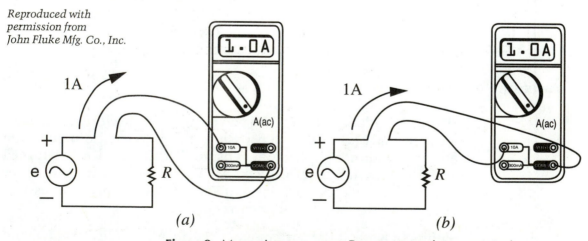

Figure 8   Measuring ac current. Be sure to use the current jack.

How To Measure Resistance

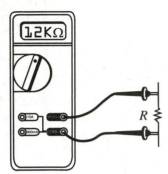

Figure 9 shows a DMM connected to measure the resistance of an isolated resistor. 1) Set the dial to resistance; 2) Connect the probe across the component whose resistance you wish to measure; 3) Read the measured value. Be sure to note the unit, Ω, kΩ, etc. (Further details may be found in Lab 2.)

Figure 9 Measuring resistance. Reproduced with permission from John Fluke Mfg. Co., Inc.

Interpreting Analog Meter Scales

Analog instruments have multiple scales, and values are somewhat more difficult to interpret than for digital instruments. To interpret a result, you must note both the scale reading and the range to which the function selector switch is set. For example, in Figure 10, the dial

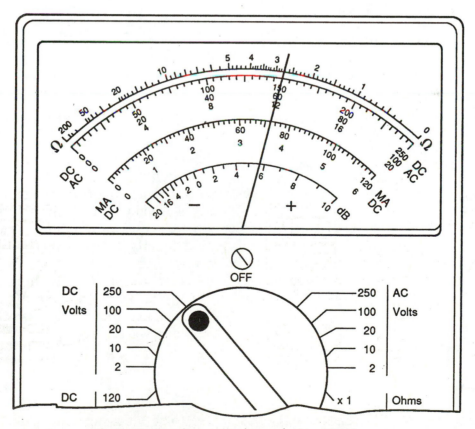

Figure 10 Reading an analog scale. V = 150 V.

is set to the 250-V range and you must therefore read the voltage on the 250-V scale. Here, V = 150 V.

Note that analog scales are divided into intervals (called *primary divisions*) with unmarked graduations (called *secondary divisions*) between. For example, on the 250-V range of Figure 10, each primary division represents 50 V. By counting the number of secondary divisions between 50-V markings, you can see that each represents an increment of 5 volts. Reading a value is straightforward if the needle is over a primary division as indicated in Figure 10. Reading a voltage is also straightforward if the needle rests over a secondary division as in Figure 11(a). By counting the number of 5 V increments, you can see that the meter indicates 160 V. However, if the needle rests part way between graduations, you must estimate the reading. For example, the reading shown in Figure 11(b) might be interpreted by one person as 183 V but by another person as 184 V. Thus, the estimated digit will have some uncertainty.

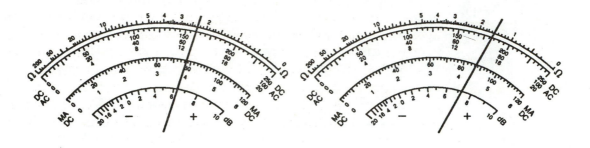

(a) V = 160 V.   (b) *There is some uncertainty here.*

Figure 11 Interpreting meter scales. The meter here is set to the 250-V range.

Range Selection

As noted, some scales are used by more than one range. For example, in Figure 10 the 100-V scale is used by both the 100-V and the 10-V ranges. When the selector switch is set to 100-V, full-scale voltage is 100 volts and the scale is read directly, with each secondary division representing 2 volts. However, when it is set to 10-V, full-scale voltage is 10 volts and readings must be divided by 10. In this case, secondary divisions represent 0.2 volts. Thus, for Figure 10, if the range selector were set to 100-V, the reading would be 60 V, while if it were set to 10-V, the reading would be 6 volts.

Meter Errors

No meter can be guaranteed to be 100% accurate. Thus, when you measure a quantity, there will always be some uncertainty due to the meter itself. (This is similar to the situation with bathroom scales. If your scale is not perfectly accurate, it may indicate a few pounds higher or lower than your actual weight.) In general, DMMs are more accurate than analog VOMs.

Meter Accuracy Specifications

The *accuracy specification* for a meter defines the maximum error that it may have—that is, the guaranteed maximum value by which the measured value may deviate from the true value. Accuracy specifications for digital and analog meters are defined differently. For digital instruments, the maximum error is stated in terms of the actual meter reading, while for analog instruments, it is stated in terms of the meter's full scale value. This is an important consideration when estimating errors in measurement.

Digital Meters The basic dc voltage accuracy specifications for bench/portable and hand-held DMMs range from about 0.05% to 0.5%, depending on make and model. This specification is independent of reading—thus, a DMM with a 0.5% error specification will have no more than 0.5% error no matter what value it is reading. For example, if its displayed value is 100.0 volts, it could have up to 0.5% × 100 volts = 0.5 V error, meaning that the true value of the measured voltage can be anywhere between 99.5 and 100.5 volts. The same meter indicating 16.00 volts still has a possible 0.5% error of its reading (which in this case is 0.5% × 16 volts = 0.08 V) and thus, its true value could be anywhere between 15.92 and 16.08 volts.

Some DMMs also have an additional uncertainty in their least significant digit. For example, a certain popular low cost, hand-held DMM has a specification of ±(0.5%+1 digit) which means that its maximum error at any point will not exceed one half a percent of its indicated reading plus one least significant digit. Thus if this meter is displaying 100.0 volts, it could have an error of half a volt (as noted above) plus 1 digit. This means that the true value of the measured voltage lies somewhere between 99.4 and 100.6 volts.

VOM The accuracy of a VOM is defined in terms of its full scale value. Thus a VOM with an accuracy specification of 2% measuring a voltage on a 50-V scale may have as much as 1 volt error anywhere on that scale (since 2% of 50 V is 1 V). While this 1 volt error represents 2% at full scale, it represents 4% at half scale and 10% at the 10-V point. Thus, for a measured value of 50 volts, the true value may lie anywhere between 49 and 51 volts, but for a measured value of 10 volts, the true value may be anywhere between 9 and 11 volts. Accuracy can be improved by changing to a range where the reading is closer to full scale. For example, a 2% error on the 10-V range represents a possible error of 0.2 volts; thus, if you measure 10 volts on the 10-V range, the true value lies between 9.8 and 10.2 volts, a considerable improvement over the same reading taken on the 50-V scale. (It is for this reason that you should select a range that yields a reading close to full scale when using a VOM.) In addition, VOM accuracies are usually specified with the meter lying in a horizontal position; if the meter is vertical, its accuracy could be poorer due to the construction of the needle pivot. It may help to tap the meter lightly if the needle appears stuck.

A Final Note

Accuracy specifications for ac measurements are generally poorer than for dc measurements. (Check the operator's manual for your particular meter for details.) Note also that accuracy specifications define a meter's worst case error and that most meters will perform better than their published accuracy specifications for many measurements. Still, the published figure is the only guarantee that you get. Note also that manufacturers may guarantee their specifications for a limited time only (e.g., one year) after calibration.

Schematic Symbols for Meters

So far, we have shown meters in pictorial form so that you could see how the meters were physically connected into a circuit and how the function selector switch is used. In practice, this is cumbersome and meters are usually represented by simple schematic symbols as in Figure 12. (In most labs, we show only the schematic. However, in the first few labs, we sometimes show both.)

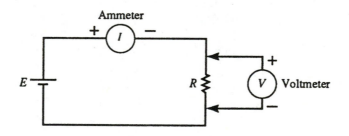

Figure 12   Schematic symbols

Words of Caution

1. Never connect an ammeter across a voltage source. An ammeter is virtually a short circuit and damage to the meter or the source may occur.
2. Always plug the probe leads into the meter sockets before connecting the probe tips to the circuit.

3. When using a meter with manual range selection, set the meter to its highest range if you do not know the approximate value of the quantity to be measured, then switch to lower ranges until you get to a suitable range.

The Function Generator

A *function generator*, Figure 13, is a variable frequency, multi-waveform source that produces a variety of waveforms such as sine, square, triangular, pulse, and so on. It will have a set of push-buttons or a selector switch for selecting the desired waveform, range selector push-buttons or a selector switch for setting the frequency range, a variable control for adjusting the frequency within the selected range, an amplitude control for adjusting the output voltage, and a digital readout (or an analog scale on the dial) for displaying the chosen frequency. A *dc offset* control is usually included so that you can add a positive or negative dc voltage component to the waveform. Depending on the make and model, other features may also be included. The usual frequency range of a function generator is from a few Hz to a few MHz.

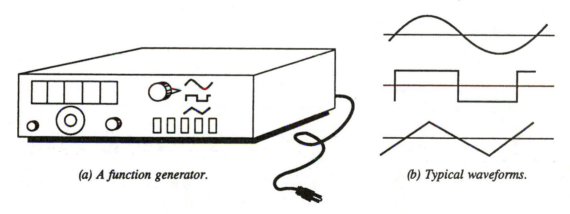

(a) A function generator.   (b) Typical waveforms.

Figure 13   A function generator and some of its waveforms

The Oscilloscope

The *oscilloscope* is an electronic test and measurement instrument that displays waveforms on a screen. It is the basic tool used for studying time-varying phenomena such as voltages and currents in electric circuits. It is used, for example, to measure the frequency and period of repetitive waveforms, to determine the rise and fall times of pulses, to find the phase difference between sinusoidal signals, to help troubleshoot electronic equipment, and so on. However, the oscilloscope is a rather complex instrument. We will therefore look at it in several stages. In this section, we look at it from a conceptual viewpoint; in later sections, we learn how to use it to make voltage, current, phase angle, and other measurements in a lab setting. Figure 14 shows a typical dual-channel oscilloscope.

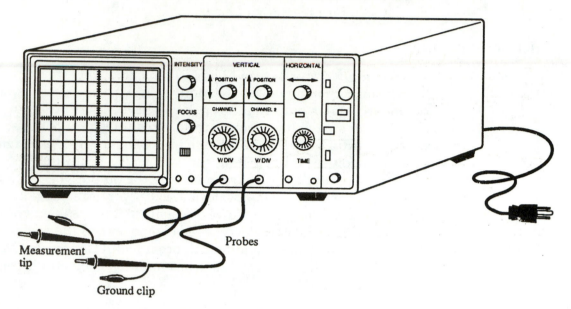

Figure 14   A typical dual-channel oscilloscope. Oscilloscopes are also referred to as scopes.

The Display Screen

The display screen of a standard oscilloscope is 10 cm (about four inches) across and is ruled with a grid at 1 cm intervals with graduated markings (called *graticules*) between grid lines. Measurements are made by positioning waveforms on this grid and reading values from its scale as illustrated in Figure 15. By selecting appropriate vertical and horizontal scale factors, you can determine the amplitudes of the waveforms, their period, and the phase displacement between them.

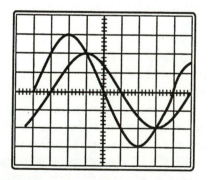

Figure 15   Waveforms on the screen

Inside the Oscilloscope

Figure 16 is a simplified block diagram of an oscilloscope. It consists of a *CRT* (cathode ray tube), a *vertical display section*, a *horizontal display section* and *trigger circuits*.

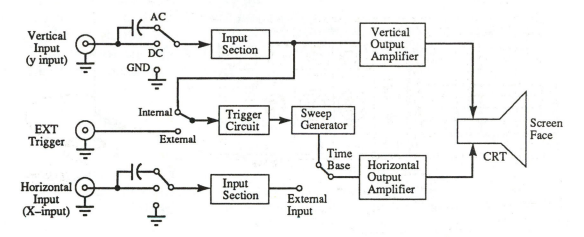

Figure 16   Simplified block diagram of an oscilloscope

The CRT

The heart of the oscilloscope is its CRT. (It displays the waveform.) The CRT is a vacuum tube, one end of which is flared out to make the screen. The screen is coated on the inside with a thin layer of phosphorescent material, and an *electron gun* inside the tube produces a stream of electrons and shoots them onto this screen. As each electron strikes the screen, it produces a spot of light. A horizontal deflection system then sweeps this spot across the face of the screen at a rapid rate, producing a horizontal line. This horizontal line can be positioned by means of a *vertical position control* located on the front panel. With no input signal present, you will usually position this line at the center of the screen.

The Vertical Deflection System

When an input signal is applied to the vertical input, the *vertical deflection system* causes the spot to move vertically in proportion to the voltage applied. With the spot moved vertically in response to the input signal and horizontally by the deflection system, a path is traced on the screen corresponding to the input waveform. If a dual-trace scope is used, two waveforms can be displayed simultaneously as in Figure 15. A vertical input is sometimes called a "y" input.

Because input signals may be large or small, an input *attenuator* and an *amplifier* are provided. A front panel control permits you to set the attenuation/amplification to suit the measurement. This control is calibrated in volts per division (VOLTS/DIV). A variable control allows for fine adjustment, but it is not calibrated. A switch on the input line allows you to select *ac coupled* or *dc coupled*. The switch also includes a GND (ground) position as indicated in Figure 16. When set to *dc*, the input signal is applied directly to the vertical deflection system, permitting the entire signal (ac and any dc component present) to be viewed on the screen; when set to *ac*, the input signal passes through a capacitor which blocks all dc and results in only the ac signal being displayed. The GND position permits you to establish a 0-V baseline by internally applying zero volts to the vertical deflection system.

The Horizontal Deflection System

The *horizontal deflection system* may be operated in one of two modes. In the *time-base* mode, an internal sweep circuit causes the beam to sweep across the screen as described above. This circuit has a front panel control that you use to set the time per division (i.e., the length of time that it takes the spot to move from one grid line to the next). For example, if you are measuring a high frequency waveform, you need to set the horizontal sweep at a high rate. Typically, this control is calibrated in seconds, ms and µs per screen division. The other mode of operation is the *horizontal input mode* which permits external signals to be applied to the horizontal input (sometimes called the "x" input). This is useful if you wish to display an x-y trace on the screen. On dual-channel oscilloscopes, one of the channels normally doubles as the x-input when used in the x-y mode.

Trigger Controls

To display a periodic waveform, the spot is swept repeatedly across the screen. The job of the *trigger circuit* is to always start the trace at the same point on the waveform so that each succeeding trace is superimposed on the preceding one, making it appear stationary on the screen. The trigger circuit synchronizes this process. As indicated in Figure 16, it samples the vertical input to determine when to start the trace moving horizontally across the screen. Two front panel controls, the *trigger level* and the *trigger slope* control permit you to select the trigger point. If triggering is not set properly, the trace will not be stable on the screen.

Sophisticated oscilloscopes have additional triggering controls. We will not consider them here.

Other Controls

Other controls include a *focus control* and an *intensity control*. These are used to adjust the trace for a sharp, crisp trace at a comfortable viewing level. A *horizontal position control* permits you to adjust the trace horizontally. A *beam finder* helps you locate the trace if it is off the screen and you are having trouble finding it.

Frequency Range

Oscilloscopes have an extremely broad frequency range; even low-cost models can measure signals from dc to 20 MHz. Top of the line models can measure signals up to several hundred MHz and digital oscilloscopes can measure up into the GHz range.

Probes

Probes are an integral part of a scope's measurement system. Probes are either "× 1" or "× 10" (times one or times ten). A "× 1" probe has no attenuation and thus, the full voltage at its tip is applied to the scope. A "× 10" probe, on the other hand, has a built-in ten-to-one voltage divider, and thus, only one tenth of the measured voltage is applied to the scope. With a "× 1" probe, the input resistance of the scope is 1 megohm; with a "× 10" probe, it is 10 megohms. Thus, the "× 10" probe results in less loading of high impedance circuits.

Each probe has a built-in *compensation circuit* which must be "tuned" to match the channel to which it is connected. Depending on the manufacturer, this is done by twisting the barrel of the probe or by adjusting a compensator screw on the probe. To tune the compensation circuit, touch the probe to the calibration test point on the front panel of the scope and adjust until the waveform seen on the screen is a flat-topped square wave. Once the probe is tuned, it will not introduce distortions into the measurements that you are making.

Final Comments

Although all oscilloscopes have the above basic features in common, there are differences in detail that you will have to learn from your instructor and from the scope's manual. Much of this you will learn as you perform the labs.

Safety

Everyone has handled electrical appliances and tools in one form or another and most have never experienced a serious injury or mishap involving electricity. As a result, most of us think of electricity as a safe form of energy, if we think of it at all! However, we have all heard of someone getting a nasty shock as the result of carelessness or from poorly maintained equipment.

Although it is impossible to foresee every possible injury, most electrical accidents are preventable by using common sense and by adopting a heathy respect for electrical energy. A knowledgeable technician or technologist not only avoids electrical hazards but also knows what to do when someone else has had a mishap involving electricity.

Current kills!

The human body uses electrical impulses as low as 10 mV (ten one-thousandths of a volt) to transmit nerve messages between the brain and the various parts of the body. When an external source of electricity interacts with the body, the results can be disastrous! The human body consists mostly of water, with numerous dissolved electrolytes such as potassium chloride, phosphates and sodium chloride, making us very good conductors of electricity. An alternating current as low as 5 mA (five one-thousandths of an ampere) can be quite painful, causing muscles to go into spasms. When the current is increased to 10 mA, the spasms may be sufficient to prevent the victim from letting go of the current source. If the current is increased to 15 mA, breathing may be stopped and the heart itself may go into spasms (fibrillation) preventing the flow of blood to the brain. If the victim does not receive immediate attention, death may result. The following steps must be followed as quickly as possible.

1. **Remove the victim from the source of electricity**. Since touching the victim may result in electrocution of the rescuer, it is necessary to pull or push the victim from the source of electricity without coming into electrical contact. An insulated object such as a broom or a dry stick may be applied at the midsection of the victim to push him/her from the electrical source.

2. **Call for assistance**. If another person is nearby, get him/her to telephone for medical help. If no one is nearby and a telephone is handy, call for assistance. **Do not leave the victim unattended.**

3. **Check for breathing and a pulse**. Once the victim has been removed from the electrical source, determine the immediate condition of the person. This is done by first calling his/her name. If there is no response, observe the chest or listen at the mouth for signs of breathing. If there is no sign of breathing, determine whether a heartbeat is present. This is done by placing the index and the middle fingers on the carotid artery which is located on the throat, just under the jaw and slightly ahead of the ear. If there is no pulse, it is necessary to begin cardiopulminary resuscitation (CPR) as soon as possible. If a pulse is present, but breathing has ceased, then artificial respiration (AR) is necessary. The sooner first aid is begun, the more likely it is the person will recover.

4. **Wait until professional help arrives**. If the victim feels cold and clammy, keep him or her warm with blankets or clothing. If the victim feels hot and sweaty, remove any tight clothing and give

the person fresh air. You will need to tell the attending paramedics how the victim arrived at his/her injuries and what first aid was administered.

Preventing Electrical Injury

Rather than treating an electrical injury, it best to prevent the occurance in the first place. The following is a list of safety rules which should be observed in the workplace or the lab.

1. **Consider all voltages above 50 Vdc or 30 Vac (rms) as a potential hazard**. If a voltage source is touched with dry hands you may experience only a mild tingle. The same source may be painful or fatal if your skin is moist or has a cut. If you experience even a minor electrical shock, mention this to your instructor or lab supervisor so that no one else is subjected to a potential hazard. It may be necessary to service the equipment.

2. **Ensure that all equipment is operating correctly and that no power cords are frayed or cut**. If a power cord is not in perfect condition, label the fault and bring it to the attention of your instructor or supervisor. Do not put a faulty piece of equipment into service. Tag the equipment and submit it for repair.

3. **Carefully insert and remove power cords**. Remove power cords by pulling on the connector. Never remove a power cord from the socket by pulling on the cable. There is a good chance that the cable will eventually tear free of the connector, presenting a fire or electrical hazard.

4. **Consider all sources of RF** (radio frequency) **energy as potential hazards**. Signals at very high frequencies (such as microwaves) have the potential to damage molecular structures (including your body's skin and internal organs) without immediately apparent symptoms.

5. **Use electrolytic capacitors with caution**. Some electrolytic capacitors are able to store large amounts of charge and may have the capacity to inflict the same injuries as voltage sources. Ensure that electrolytic capacitors are connected into the circuit with the correct polarity and verify that the voltage rating of the capacitor is not exceeded. If an electrolytic capacitor is connected incorrectly, the result may be an explosion, sending electrolytic chemicals and sharp pieces of metal into the air. Safety glasses provide some protection from shrapnel and chemical spills.

6. **Use the correct protective equipment**. Prior to working on circuits with dangerous voltages, interrupt the circuit at the control panel by opening the switch. To prevent someone from inadvertently closing the switch, tag the circuit at the control panel, clearly stating the reason for the interrupt. Wear the appropriate protective gear such as safety glasses, gloves and boots. Avoid

wearing jewelry and loose fitting clothing. Use a rubber mat and rubber-soled boots to work on circuits in a damp environment.

7. **Know the location of fire extinguishers and circuit breakers**. In the event of an accident, the power should be immediately turned off. If a fire has broken out, you may need to use a fire extinguisher. Ensure that the extinguisher is appropriate for the particular application. A fire should only be fought if it is not between you and a safe exit and if there is a good chance that you can bring the fire under control.

8. **Follow your instructor's safety instructions**. Since each lab has its own procedures which need to be followed, your instructor (or lab supervisor) will provide you with special precautions for the equipment or components used.

A Note Concerning Calculated and Measured Results

In the real world, calculated results and measured results seldom agree exactly. It is important to understand that there are usually explainable reasons for the differences, such as component tolerances, inherent meter errors, human measurement errors and so on. As you analyze the data for each experiment, you should try to determine the explainable differences, and discuss them as part of the analysis for the lab.

Trademarks

OrCAD Capture and *OrCAD PSpice* are registered trademarks of OrCAD, Inc.

Powerstat® is a registered trademark of The Superior Electric Co.

LAB
1

Introduction to Meters and Voltage Measurement

Objectives

After completing this lab, you will be able to
- measure voltage using digital and analog multimeters,
- determine accuracy of a measured value for both analog and digital instruments,
- explain polarity as it pertains to voltage measurement,
- set the zero adjust of an analog multimeter,
- read analog meters with minimal parallax error.

Equipment Required

☐ Digital multimeter (DMM)
☐ Volt-ohm-milliammeter (VOM)
☐ Variable dc power supply
☐ 9-V battery

PRELIMINARY

Before you start, review the section on meters and dc voltage measurement in *A Guide to Lab Equipment and Laboratory Measurements* at the front of this manual.

EQUIPMENT USED

Instrument	Manufacturer/Model No.	Serial No.
DMM		
VOM		
Power supply		

Table 1-1

TEXT REFERENCE

Section 2.6   MEASURING VOLTAGE AND CURRENT

DISCUSSION

In principle, measuring voltage is easy—you simply connect a voltmeter across the circuit element whose voltage you wish to determine, then read the value from the meter as indicated in Figure 1-1. However, there are important practical considerations. In this lab, you will look at a number of these; you will, for example, learn what meter connections to use when making a voltage measurement, how to select appropriate ranges, how to interpret polarity, how to read complex VOM scales, and so on. Learning how to use meters and make measurements correctly is an important skill. We begin with a look at digital multimeters and voltage measurements.

MEASUREMENTS

PART A: Measuring dc Voltage with a DMM

DC voltage measurements are made as illustrated in Figure 1-2. (Although a hand-held DMM is illustrated, the connection is the same for a bench DMM.) Note that color coded leads are used, red for the

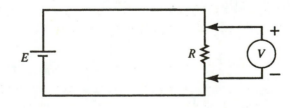

Figure 1-1   Measuring voltage in a dc circuit

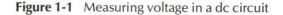

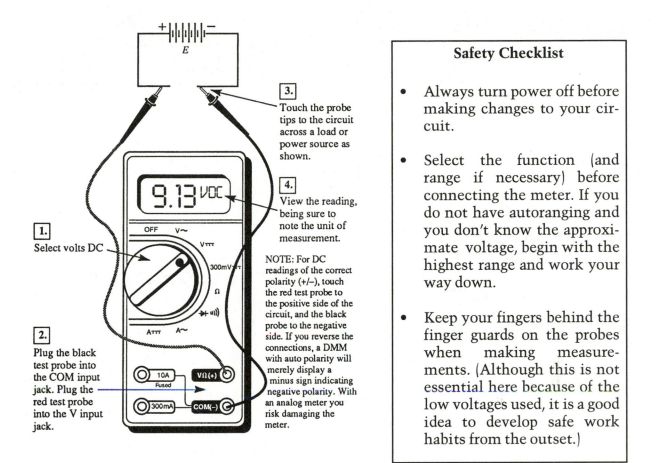

Safety Checklist

- Always turn power off before making changes to your circuit.

- Select the function (and range if necessary) before connecting the meter. If you do not have autoranging and you don't know the approximate voltage, begin with the highest range and work your way down.

- Keep your fingers behind the finger guards on the probes when making measurements. (Although this is not essential here because of the low voltages used, it is a good idea to develop safe work habits from the outset.)

Figure 1-2 Making a dc voltage measurement with a DMM. Steps 1 and 2 may be reversed. (Adapted from The ABCs of DMMs. Courtesy of John Fluke Mfg. Co., Inc. Used with permission.)

$V\Omega$ or (+) lead and black for the *COM* lead. (Check your meter—as noted earlier, if it uses a combined voltage/resistance/current input, the jack will be designated *VΩA* instead of *VΩ*.)

1. Connect the probes to the DMM as in Figure 1-2, i.e., the red probe to the VΩ (or VΩA) jack and the black probe to the COM jack. (If your meter has alternate designations, check with your instructor.) Select *Volts DC* and set to the appropriate range if your meter is not autoranging. (20 V is adequate for this test.) Connect the DMM to the variable power supply. Adjust to 10 volts. Note the polarity of the meter reading.

    Meter reading _____

2. Reverse the test probes and note the new reading, including its sign.

    Meter reading _____

3. Repeat tests 1 and 2 at $E = 0.5$ volts. Indicate meter readings and your observations concerning range changing, i.e., did your meter automatically change range or did you have to do it manually?

_____

_____

4. Add the 9-V battery to the circuit as in Figure 1-3. Take voltage readings with E set to 0 V, 5 V, 10 V, and 15 V and record data in Table 1-2. Based on these results, write an equation for V in terms of E and the 9-V battery.

$V =$

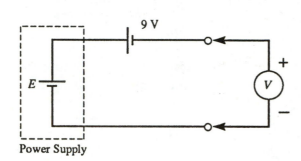

Power Supply

Figure 1-3   Circuit for Step 4, a series opposing circuit

E	V
0 V	
5 V	
10 V	
15 V	

Table 1-2

5. Reverse the 9-V battery as in Figure 1-4 and repeat Step 4. Record data in Table 1-3.

$V =$

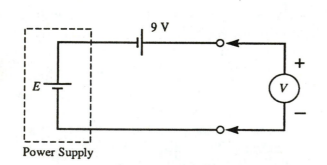

Power Supply

Figure 1-4   Circuit for Step 5, a series aiding circuit

E	V
0 V	
5 V	
10 V	
15 V	

Table 1-3

PART B: Measuring dc Voltage with a VOM

DC voltage measurements using a VOM are made as illustrated in Figure 1-5. As with the DMM, the meter will indicate a positive (or

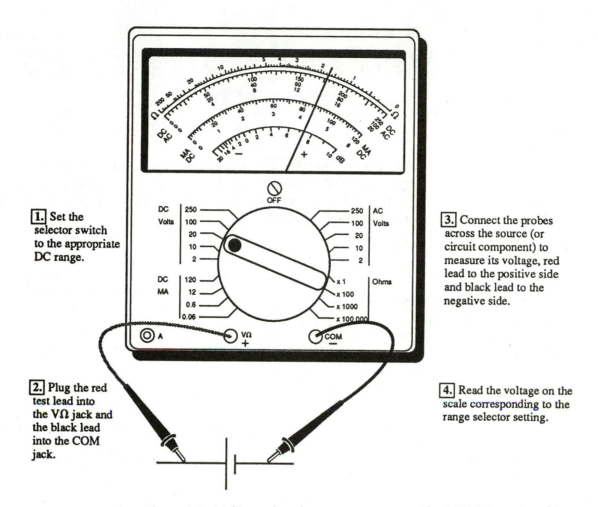

1. Set the selector switch to the appropriate DC range.

2. Plug the red test lead into the VΩ jack and the black lead into the COM jack.

3. Connect the probes across the source (or circuit component) to measure its voltage, red lead to the positive side and black lead to the negative side.

4. Read the voltage on the scale corresponding to the range selector setting.

Figure 1-5   Making a dc voltage measurement with a VOM. Steps 1 and 2 may be reversed.

upscale) value when the VΩ lead is connected to the positive side of the circuit and the COM lead is connected to the negative side. Unlike the DMM, however, *you cannot interchange leads; if the leads are reversed, the meter will attempt to read down scale, and you risk damaging the instrument.*

Zeroing a VOM

With no voltage applied, a VOM should indicate zero. However, analog meters sometimes go out of adjustment. (You should always check the zero position of your meter before you make any measurements.) To set the pointer back to zero, disconnect the meter from the circuit, place the meter in its operating position, then turn its *zero adjust screw* until the pointer rests exactly over the zero mark on the scale. Now carefully reverse the screw a slight amount to introduce some *play* into the system. However, be careful not to disturb the pointer as you do this—it should remain at zero throughout the process.

Parallax Errors

Parallax results when you view a meter at other than at right angles to its face. To illustrate, consider Figure 1-6. Assume the pointer is exactly over the 150 V mark. When viewed straight on as in (a), you read 150 V; when viewed slightly to one side as in (b) or (c), you get a small error. To avoid parallax, some meters use a *mirror-backed scale*. When you align the pointer with its reflection, you reduce the possibility of parallax error.

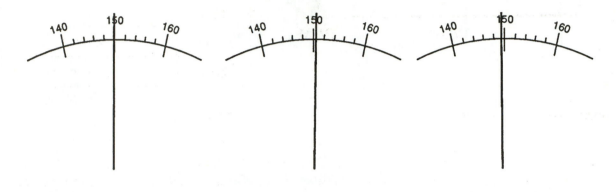

(a) No parallax error.   *(b) Parallax error.*   *(c) Parallax error.*

Figure 1-6   In all cases, the pointer is exactly over the 150 V mark.

MEASUREMENTS

If you have trouble interpreting VOM readings, review the section on analog scales in *A Guide to Lab Equipment and Laboratory Measurements* at the front of this manual.

6. a. Zero your VOM as described above, then connect both it and a DMM to the power supply as in Figure 1-7.

    b. Using the VOM to measure the voltage, carefully set the power supply to 2 V. (Don't look at the DMM, as the objective here is to see how close you can come using the VOM.) When you have set the voltage to your satisfaction using the VOM, read and record both voltmeter readings in Table 1-4.

    c. Repeat test (b) for E = 10 V, 15 V, and 20 V plus some off division values such as 5.7 V, 13.1 V, and 17.6 V.

    d. With a calibrated and properly zeroed VOM, you should have been able to set the supply accurately to the voltages that correspond to primary and secondary scale markings. However, if you are like most people, you probably didn't do so well for the off division values. Now repeat, using the DMM to set the volt-

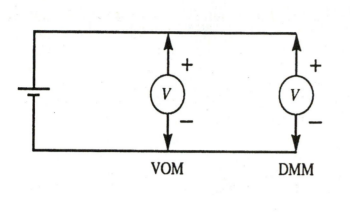

E	VOM	DMM
2 V		
5.7 V		
10 V		
13.1 V		
15 V		
17.6 V		
20 V		

Figure 1-7 Use both meters.

Table 1-4

ages. Note how easy it is. This is of course one of the reasons why the DMM has largely superseded the VOM in practice.

PART C: Accuracy of Measured Values

As noted earlier in *A Guide to Lab Equipment and Laboratory Measurements*, no meter can be guaranteed 100% accurate. However, DMMs are considerably better than VOMs. Recall:

- VOMs are rated in terms of full scale. A typical specification is $\pm 2\%$ of full scale, which means that at half scale, the error could be $\pm 4\%$, while at less than half scale, it is even worse. (This is why you should make measurements as close to full scale as possible.)
- The accuracy specification for a DMM is given in terms of its actual reading and, even for low-cost instruments, is typically $\pm(0.5\%$ plus 1 digit) or better.

Meter Ranges The VOM referenced in this lab has 2-V and 10-V ranges, etc. If your meter does not have exactly the same ranges, use the closest ones available. For example, if your meter has a 2.5-V scale instead of a 2-V scale, use it.

MEASUREMENTS

7. a. Using the circuit of Figure 1-7, set the supply to $E = 2.00$ V using the DMM.

   b. Read the voltage on the 2 V scale of the VOM. $V = $ _____

   c. Read the voltage on the 10 V scale of the VOM. $V = $ _____

d. Assuming ±0.5% possible error for the DMM and ±2% full scale for the VOM, compute between what guaranteed limits the actual voltages lie for the readings of Steps 7(a), (b) and (c).

PROBLEMS

8. What color probe should you connect to the VΩ terminal? To the COM terminal?

_____

9. Using a DMM, what happens when you connect the VΩ probe to the negative side of a circuit and the COM probe to the plus side?

_____

10. Using a VOM, what happens when you connect the VΩ probe to the negative side of a circuit and the COM probe to the plus side?

_____

11. The DMMs of Figure 1-8 have autopolarity. What does each indicate? (Show your answer in the DMM display block.)

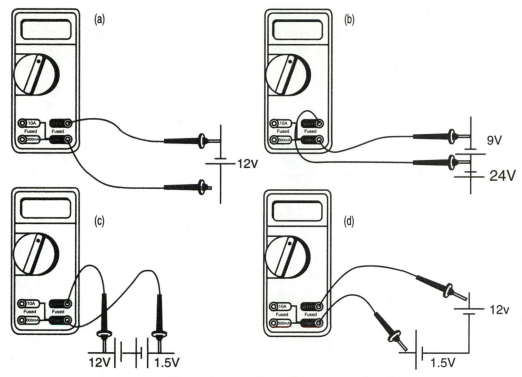

Figure 1-8. Indicate the reading of each meter.

12. For the VOM of Figure 1-9, determine the voltage indicated for each setting of the range selector shown in Table 1-5.

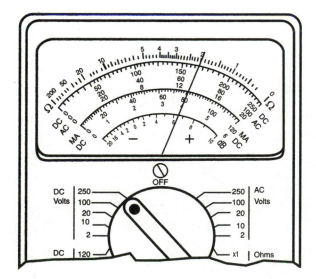

Range (V)	Value
250	
100	
20	
10	
2	

Figure 1-9 What is the meter reading?   **Table 1-5**

13. For the meter of Figure 1-10, draw in the pointer position and the range selector setting for the most accurate reading if the voltage being measured is 18 volts dc.
14. A DMM indicating 15.00 volts has an accuracy specification of ±(0.2%+1 digit). Between what two values does the true value of the measured voltage lie?

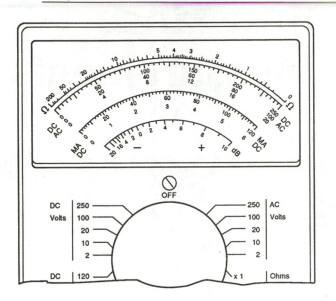

Figure 1-10 Add the pointer and range setting.

Resistance Measurements

OBJECTIVES

After completing this lab, you will be able to
- measure resistance with analog and digital ohmmeters,
- compare color band values with measured values of resistance,
- compare the physical dimensions of resistors of different power rating,
- measure the resistance of a diode,
- observe how temperature affects resistors and thermistors.

EQUIPMENT REQUIRED

☐ Digital multimeter (DMM)
☐ Volt-ohm-milliammeter (VOM)
☐ Ruler or caliper scaled in centimeters
Note: Record this equipment in Table 2-1.

COMPONENTS:

☐ resistors: 47-Ω, 220-Ω, 3.9-kΩ, 10-kΩ, 180-kΩ, 5.6-MΩ (all resistors are 1/4-W, 5% tolerance)
470-Ω (1/8-W), 470-Ω (1/4-W), 470-Ω (1/2-W), 470-Ω (1-W), 470-Ω (2-W).
☐ thermistor: 10-kΩ @ 25°C (or equivalent)
☐ diode: 1N4004 (or equivalent)

EQUIPMENT USED

Instrument	Manufacturer/Model No.	Serial No.
DMM		
VOM		
Other		

Table 2-1

TEXT REFERENCE

Section 3.6 COLOR CODING OF RESISTORS
Section 3.7 MEASURING RESISTANCE—THE OHMMETER
Section 3.8 THERMISTORS
Section 3.9 NON-LINEAR RESISTANCE

DISCUSSION

Fixed resistors are used extensively in electrical and electronic circuits to limit the amount of current in the circuits. Although the value of a resistor (in ohms) is occasionally stamped on the resistor (e.g., 330[330-Ω], 2k0[2.0-kΩ]), it is more common to designate the resistance value with a standardized color code. Figure 2-1 shows a typical color-coded resistor.

The procedure for determining the value of a color-coded resistor is described in Section 3.6 of the textbook and is briefly outlined in Appendix A. Resistor color codes are read from left to right, left being defined as the end of the resistor with the band nearest to the end. The first two bands of the resistor provide the first and second digits of the resistance value. The third band is the multiplier and represents the number of zeros following the first two digits. The fourth band provides the tolerance (or uncertainty) of the resistance value. A fifth band is occasionally used to represent the reliability of the resistor. This last band is used primarily in military applications and seldom used in commercial electronics. The physical size of the resistor is an indication of how much power the resistor can safely dissipate. Resistors with a large surface area can handle more power than small resistors.

When measuring resistance with an ohmmeter, several important steps must be followed.

1. All power supplies must be disconnected from the circuit. If power is not removed, the ohmmeter may be damaged.

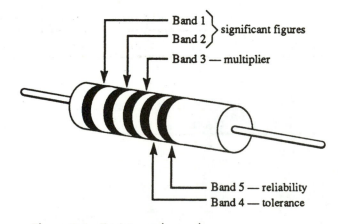

Band 1 ⎫
Band 2 ⎬ significant figures

Band 3 — multiplier

Band 5 — reliability
Band 4 — tolerance

Figure 2-1 Resistor color codes

2. If the component is part of a larger circuit, it is necessary to isolate the component from the rest of the circuit. This is done by disconnecting at least one terminal of the component from the circuit.

3. Connect the two probes of the ohmmeter across the component to be measured. Be careful not to touch both probes of the meter, since your body's resistance may introduce additional error. When measuring resistance, the red and black leads of the ohmmeter can usually be interchanged without affecting the reading. The exception is diodes and most active components such as transistors and integrated circuits (ICs).

4. Select a range which is appropriate for the measurement to be taken. Prior to taking measurements a meter movement may need to be zeroed. This is generally done by adjusting a ZERO OHMS potentiometer. When using a meter with a PMMC (permanent magnet moving coil) movement (such as a VOM), the most accurate reading is obtained when the needle is approximately midscale. You must remember to multiply the reading by the corresponding scale multiplier. For example, a reading of 2.4 on the R×100-Ω range corresponds to a resistance of 240 Ω. When using a digital ohmmeter, the ranges represent the largest resistance which can be measured on the scale. For example, a reading of 2.4 on the 200-Ω range corresponds to a resistance of 2.4 Ω, while the same reading on the 20-kΩ range corresponds to a resistance of 2.4 kΩ.

5. When storing the ohmmeter, it is necessary to ensure that the ohmmeter is turned off, since it is possible to drain the internal battery if the probes are accidently connected together for any length of time.

CALCULATIONS

1. Refer to the color coded resistors given by your lab instructor. Determine the resistor value and the corresponding tolerance for each resistor and enter your results in Table 2-2.

	Color Codes	Resistance and Tolerance
Example	Brown Black Orange Gold	10 kΩ ± 5%
R_1		
R_2		
R_3		
R_4		
R_5		
R_6		

Table 2-2

2. The tolerance of the resistor is a specification indicating the range of possible values in which the actual resistance of the component will occur. The resistor in the example of Table 2-2 has a nominal value of 10 kΩ with a tolerance determined as follows:

$$\text{Tolerance} = (5\%)(10 \text{ k}\Omega) = 0.5 \text{ k}\Omega = 500 \text{ }\Omega$$

By specifying a tolerance of 5%, the manufacturer of the resistor is providing assurance that the actual value of the resistor is between 9.5 kΩ and 10.5 kΩ. Calculate the expected minimum and maximum values for each of the resistors in Table 2-2. Enter the data in Table 2-3.

	Minimum Resistance	Maximum Resistance
Example	9.5 kΩ	10.5 kΩ
R_1		
R_2		
R_3		
R_4		
R_5		
R_6		

Table 2-3

MEASUREMENTS

3. Connect the ohmmeters as shown in Figure 2-2 and measure each of the resistors. Record the results in Table 2-4. Reverse the leads of the ohmmeter and measure again. There should be no difference in the reading.

	DMM	VOM	Other
Example	9.83 kΩ	9.8 kΩ	9.90 kΩ
R_1			
R_2			
R_3			
R_4			
R_5			
R_6			

Table 2-4

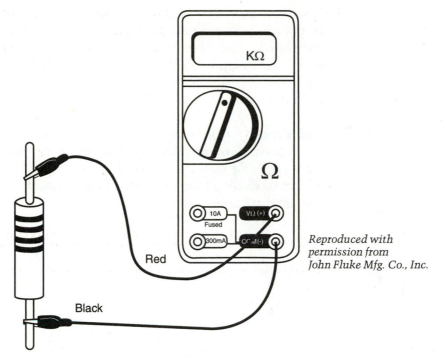

Figure 2-2   Measuring resistance with an ohmmeter

4. Select five resistors with the same color bands but different power ratings. Record the color codes and expected resistance values for the resistors on the top portion of Table 2-5. Since the codes are the same, the resistors will all have the same value. Use the DMM to measure each of the five resistors. Enter the results in the appropriate column of Table 2-5.

Resistor Color Code	Resistance and Tolerance

	Measured Resistance	Volume
R_1 (1/8-W)		
R_2 (1/4-W)		
R_3 (1/2-W)		
R_4 (1-W)		
R_5 (2-W)		

Table 2-5

5. Use a ruler or caliper to measure the physical dimensions (length and diameter) of the resistors. Calculate and record the results in Table 2-5.

Diodes are electronic devices which permit current in one direction, while preventing current in the opposite direction. The principle is the same as a check valve in a water line. Although the theory of operation is outside the scope of this book, we will observe how the resistance of a diode is affected by the way it is connected into a circuit.

6. Connect the VOM to the diode as illustrated in Figure 2-3(a). The diode is said to be operating in its forward region. Record the results in Table 2-6.
7. Connect the VOM to the diode as illustrated in Figure 2-3(b). The diode is said to be operating in its reverse region. Record the results in Table 2-6.
8. Repeat the diode measurements using a DMM. It may be necessary to move the range switch to the "diode" position (if one is available). Record the results in Table 2-6.

	VOM		DMM	
	Forward	Reverse	Forward	Reverse
Diode				

Table 2-6

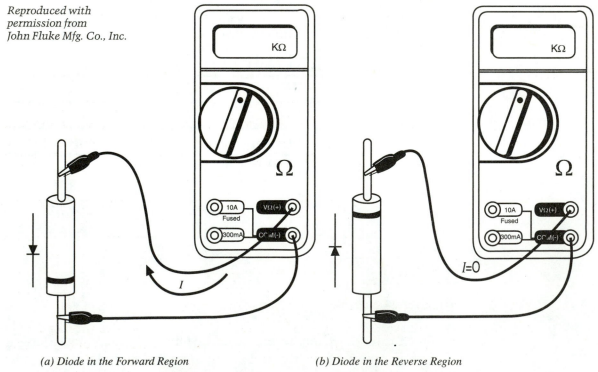

Reproduced with permission from John Fluke Mfg. Co., Inc.

(a) Diode in the Forward Region   *(b) Diode in the Reverse Region*

Figure 2-3   Measuring diode resistance with an ohmmeter

A thermistor is a *transducer* in which the resistance changes with changes in temperature. If the resistance decreases as the temperature increases, the component is said to have a *negative temperature coefficient* (NTC). Conversely, if the resistance increases as the temperature increases, the component is said to have a *positive temperature coefficient* (PTC).

9. Connect a DMM to the terminals of a thermistor. Measure the resistance of the thermistor at room temperature and record the result in Table 2-7. With the DMM still connected to the thermistor, hold the body of the thermistor with two fingers. The additional temperature from your fingers will raise the temperature of the component to approximately 30° C. Once the meter display has settled, record the approximate resistance of the thermistor in Table 2-7.

10. Repeat the measurement of Step 9 with a resistor having a similar resistance. You should observe that the resistance remains relatively constant in comparison to the thermistor. Record the results in Table 2-7.

	Thermistor	Resistor
T = 20°C		
T = 30°C		

Table 2-7

CONCLUSIONS

11. Compare the resistance measurements of Table 2-4 to the expected tolerances in Table 2-3. Are the measured values of all resistors within the tolerances specified by the colored bands? If not, indicate which resistors fall outside the accepted values and calculate the actual percent deviation of the measured values from the color code values.

_____

_____

_____

12. Refer to the data of Table 2-5.

a. Are all the resistor values within the specified tolerance? If not, indicate which resistors fall outside the accepted values.

_____

b. What conclusion can you make about the power rating of a resistor and its physical size?

_____

13. Compare the forward resistance of a diode to its reverse resistance. Do both meters provide the same readings?

_____

_____

_____

14. Compare the resistance of the thermistor at room temperature to the reading when the temperature was increased by the heat in your fingers. Does the thermistor have a negative temperature coefficient or a positive temperature coefficient?

_____

_____

15. Compare the resistance of a resistor at room temperature to the reading when the temperature was increased by the heat in your fingers. Does the resistor have a negative temperature coefficient, positive temperature coefficient, or is resistance constant?

_____

_____

PROBLEMS

16. Complete Table 2-8 by giving the resistance value of each resistor and the corresponding tolerance (in percent).

	Color Codes	Resistance and Tolerance
R_1	Brown Red Yellow Gold	
R_2	Orange White Gold Gold	
R_3	Blue Gray Green Silver	
R_4	Yellow Orange Brown Silver	
R_5	Brown Green Orange	

Table 2-8

17. Explain how an ohmmeter could be used to determine whether a light bulb is burned out.

_____

_____

18. A VOM is used to measure the resistance of a diode. In one direction, the resistance of the diode is measured to be 250 Ω while in the other direction, the resistance is measured to be 380 Ω. Is this diode faulty? Explain your answer.

_____

_____

_____

Name _____

Date _____

Class _____

Ohm's Law

OBJECTIVES

After completing this lab, you will be able to
- measure current in a dc circuit,
- confirm Ohm's law by direct measurement.

EQUIPMENT REQUIRED

☐ Two multimeters (DMM or VOM. DMMs recommended.)
☐ Variable dc power supply

COMPONENTS

☐ Resistors: One 1-kΩ resistors, 1/4-W, 5% tolerance or better
                  Two 2-kΩ resistors, 1/4-W, 5% tolerance or
                    better

PRELIMINARY

Before you start, review the section on measuring current in *A Guide to Lab Equipment and Laboratory Measurements* at the front of this manual.

EQUIPMENT USED

Instrument	Manufacturer/Model No.	Serial No.
DMM (or VOM) #1		
DMM (or VOM) #2		
Power supply		

Table 3-1

TEXT REFERENCE

Section 4.1   OHM'S LAW

DISCUSSION

The most fundamental relationship of circuit theory is Ohm's law. Ohm's law states that in a purely resistive circuit, current is directly proportional to voltage and inversely proportional to resistance. This means that, for a circuit with fixed resistance, doubling the voltage doubles the current, while halving the voltage halves the current and so on. Conversely, for a fixed voltage, doubling the resistance halves the current, while tripling the resistance drops the current to one third and so on. This relationship may be expressed in equation form as

$$I = \frac{V}{R} \text{ (Ohm's law)} \tag{3-1}$$

where V is in volts, R is in ohms, I is in amps and reference conventions for voltage and current are as shown as in Figure 3-1. This relationship (which holds for every resistance in a circuit) is an experimental result and we will thus investigate it experimentally. First, we need to learn how to measure current.

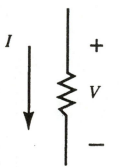

Figure 3-1   Conventions for Ohm's law.

DC CURRENT MEASUREMENT WITH A DMM OR VOM

The basic current measuring circuit is shown in Figure 3-2. For a positive or upscale reading, current must enter the appropriate current input jack and exit at COM. If in doubt, ask your instructor, as an incorrect connection may damage your instrument.

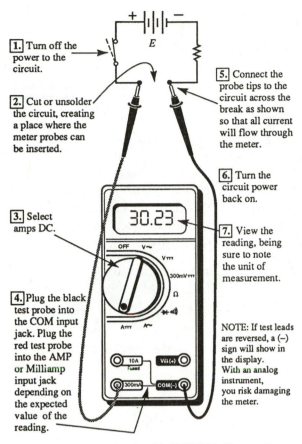

1. Turn off the power to the circuit.

2. Cut or unsolder the circuit, creating a place where the meter probes can be inserted.

3. Select amps DC.

4. Plug the black test probe into the COM input jack. Plug the red test probe into the AMP or Milliamp input jack depending on the expected value of the reading.

5. Connect the probe tips to the circuit across the break as shown so that all current will flow through the meter.

6. Turn the circuit power back on.

7. View the reading, being sure to note the unit of measurement.

NOTE: If test leads are reversed, a (−) sign will show in the display. With an analog instrument, you risk damaging the meter.

(a) Pictorial. Although a hand-held DMM is illustrated a bench-top DMM or a VOM may be used instead.

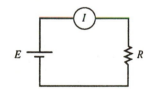

(b) Schematic

Multimeter Connections for Current Measurement

Many multimeters use a different terminal for current than for voltage. To measure current, follow the procedure indicated in Figure 3-2, connecting to the appropriate current jack as noted below.

- If your meter has a separate input jack labelled A, mA or similar, connect so that current enters the appropriate jack and exits at COM.

- If your meter has a combined voltage/current/resistance input jack VΩA, connect so that current enters here and exits at COM. Select the appropriate range.

- If your meter has a combined voltage/current/resistance input jack labelled +, connect so that current enters here and exits at COM. Select the appropriate range.

Standard Test Leads

Plug the black test lead into the COM jack and the red lead into the other jack. Connect the red lead to the positive side of the circuit and the black to the negative side as in Figure 3-2(a). This will yield a positive (or upscale) reading.

Figure 3-2   DC current measurement. (Part (a) adapted from The ABCs of DMMs. Courtesy of John Fluke Mfg. Co., Inc. Used with permission.)

Accuracy of dc Current Measurement

Depending on your meter, the accuracy of current measurements may be poorer than for voltage measurements. For example, a certain popular hand-held DMM with a dc voltage specification of $\pm(0.5\%$ of reading + 1 digit) drops to $\pm(1.5\% + 2$ digits) for current measurement. On the other hand, some meters yield the same accuracy for current as for voltage. Check your meter manual.

MEASUREMENTS

PART A: Fixed Resistance, Variable Voltage

1. Measure the value of each resistor and record in Table 3-2. (Mark each 2-kΩ resistor with masking tape so that you can keep track of them.)

	Nominal	Measured
R_1	1-kΩ	
R_2	2-kΩ	
R_3	2-kΩ	

Table 3-2

2. With power off, select R_2 and assemble the circuit as in Figure 3-3. Be sure to connect the meters so that they indicate positive or upscale values.

a. Set the voltage to $V = 8$ V and measure the current.

   $I =$ _____ (Record this result also in Table 3-3.)

b. Set the voltage to $V = 16$ V and measure the current.

   $I =$ _____

c. Set the voltage to $V = 4$ V and measure the current.

   $I =$ _____

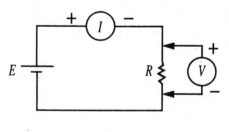

(a) Pictorial. Use DMM or VOM.   *(b) Schematic*

Figure 3-3   Connections for Ohm's law

d. Based on these observations, for a fixed resistance, how does current vary with voltage (within the limitations of accuracy of your meters and components)?

_____

PART B: Fixed Voltage, Variable Resistance

3. a. Turn off the power supply. Using both 2-kΩ resistors, connect the circuit as in Figure 3-4. (This doubles the circuit resistance to 4 kΩ.) Set V = 8 volts and measure the current.

I = _____. (Record also in Table 3-3.)

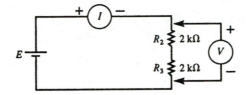

Figure 3-4   Doubling the circuit resistance. Here, R_T = 4 kΩ.

b. Turn off the power supply. Using the 1 kΩ resistor, reconnect the circuit as in Figure 3-3. Set V = 8 volts and measure the current. I = _____. (Record also in Table 3-3.)

c. Consider Table 3-3. Within the accuracy of the results obtained, for a fixed voltage, how does current vary with resistance?

| Test | $V = 8$ V | |
	R	I
2(a)	2 kΩ	
3(a)	4 kΩ	
3(b)	1 kΩ	

Table 3-3

4. For Tests 2 and 3, compute current using Ohm's law and tabulate in Table 3-4. (Use the measured values of resistance for each case.)

If there are differences between calculated and measured current, what is the likely cause? _____

_____

Test	V	R	Current I (mA)	
			Calculated	Measured
2(a)	8 V			
2(b)	16 V			
2(c)	4 V			
3(a)	8 V			
3(B)	8 V			

Table 3-4

PART C: Ohm's Law Graph

5. Using the circuit of Figure 3-3 and the 1-kΩ resistor, vary V from 0 volt to 10 volts in 2 volt increments. Tabulate your results in Table 3-5 and plot the results on the graph of Figure 3-5. Replace R with a 2-kΩ resistor and repeat.

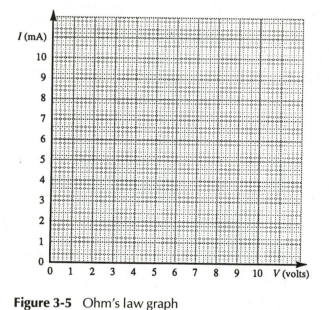

Figure 3-5 Ohm's law graph

V	Current I (mA)	
	1-kΩ	2-kΩ
0 V		
2 V		
4 V		
6 V		
8 V		
10 V		

Table 3-5

6. Resistance can be calculated at any point on the Ohm's law graph from the V and I values. For the nominal 1-kΩ resistor, at V = 5 volts, determine current from Figure 3-5 and compute R using

Equation 3-1. How well does it agree with the value of R used in Test 5? Repeat for the nominal 2-kΩ resistor.

_____

_____

_____

7. Resistance can also be computed from $R = \Delta V/\Delta I$ where ΔV represents an increment in voltage and ΔI represents the corresponding change in current. Consider the graph of Figure 3-5.

a. For the nominal 1-kΩ resistor, if $\Delta V = 4$ volts, $\Delta I =$ _____

b. Using the result of 7(a), compute the measured R. $R =$

_____

c. How does this value compare to the value of resistance determined from the color bands on the resistor?

_____

_____

PROBLEMS

8. Determine each of the unknowns.

a. $V = 6$ V, $R = 10$ kΩ. $I =$ _____

b. $R = 20$ kΩ, $I = 40$ µA. $V =$ _____

c. $V = 10$ kV, $I = 200$ mA. $R =$ _____

9. For Figure 3-3, $I = 10$ mA. If E is increased by a factor of 16 and R is halved, what is the new value of current? $I =$ _____

10. Consider the VOM of Figure 3-6 (next page). For each setting of the range selector switch of Table 3-6 and the needle position shown, determine the current reading and tabulate.

Range	Value
120 mA	
12 mA	
0.6 mA	
0.06 mA	

Table 3-6

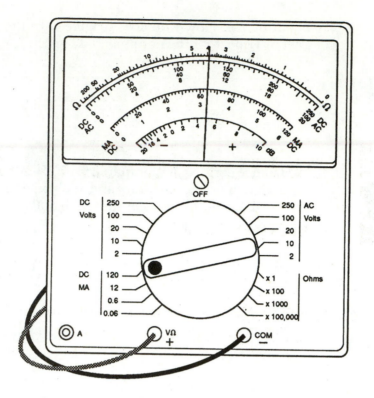

Figure 3-6   What does the meter read?

11. A current is measured as 12.00 mA using a DMM with an accuracy specification of ±(1.5% + 2 digits). What is the limit of experimental error, i.e., between what two values does the true value of current lie?

12. Ohm's law, when written in the form $I = V/R$, indicates that current I is directly proportional to voltage V. Similarly, when written in the form $V = IR$, it indicates that voltage V across a resistor is directly proportional to current I. When written in the form $R = V/I$, what does it indicate?

Series dc Circuits

OBJECTIVES

After completing this lab, you will be able to
- assemble a series circuit consisting of a voltage source and several resistors,
- use a DMM to measure voltage and current in a series circuit,
- compare measured values to theoretical calculations and verify Kirchhoff's voltage law,
- measure the effects of connecting several voltage sources in series,
- connect a circuit to ground using the ground terminal of a voltage source and use a DMM to measure voltages between several points and ground,
- measure the internal resistance of several voltage sources.

EQUIPMENT REQUIRED

☐ Digital multimeter (DMM)
☐ dc power supply (2)
  Note: Record this equipment in Table 4-1.

COMPONENTS

☐ Resistors: 47-Ω, 100-Ω, 270-Ω, 330-Ω, 470-Ω (1/4-W)
  47-Ω, 100-Ω, 270-Ω (2-W)
☐ Batteries: 1.5-V D-cell, 9-V (MN 1604 or equivalent)

EQUIPMENT USED

Instrument	Manufacturer/Model No.	Serial No.
DMM		
dc Supply		
dc Supply		

Table 4-1

TEXT REFERENCE

DISCUSSION

Two elements are said to be in *series* if they are connected at a single point and if there are no other current-carrying connections at this point. Each element in a series circuit has the same current, as illustrated in Figure 4-1.

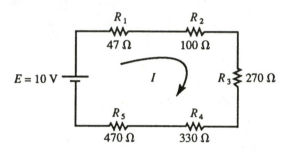

Figure 4-1   Series circuit

The equivalent resistance of n resistors in series is determined as the summation

$$R_T = R_1 + R_2 + ... + R_n \qquad (4\text{-}1)$$

When these resistors are connected in series with a voltage source, the current in the circuit is given as

$$I = \frac{E}{R_T} \qquad (4\text{-}2)$$

The voltage drop across any resisistor in a series circuit is determined by using the voltage divider rule, namely

$$V_x = \frac{R_x}{R_T}E \qquad (4\text{-}3)$$

CALCULATIONS

1. Refer to the the circuit of Figure 4-1. Calculate R_T, I, and the voltage across each resistor. Enter the results in Table 4-2. Show the correct units for each entry.

Resistor	Voltage
$R_1 = 47\ \Omega$	
$R_2 = 100\ \Omega$	
$R_3 = 270\ \Omega$	
$R_4 = 330\ \Omega$	
$R_5 = 470\ \Omega$	
R_T	
I	

Table 4-2

MEASUREMENTS

2. Connect the resistors as shown in the network of Figure 4-2. Use the DMM (ohmmeter) to measure the resistance across the open terminals. Enter the result below. Compare your measurement to the theoretical calculation recorded in Table 4-2. You should observe only a small discrepancy (no greater than the percent tolerance of the resistors.)

R_T	

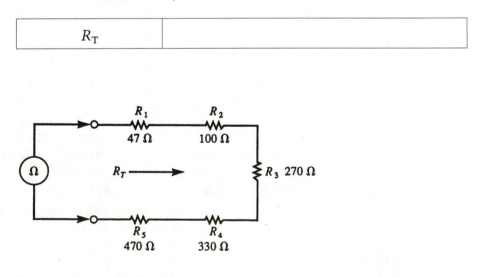

Figure 4-2   Series resistance

3. Connect the voltage source into the circuit as shown in Figure 4-3. With a DMM (voltmeter) connected across the voltage source, adjust the voltage for exactly 10 V.

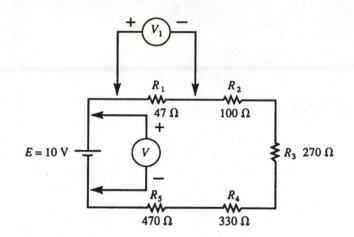

Figure 4-3   Measuring voltage in a series circuit

4. Disconnect the voltmeter from the voltage source and successively place it across each of the resistors. Measure the voltage across each resistor in the circuit and record your results in Table 4-3.

	Voltage
V_1	
V_2	
V_3	
V_4	
V_5	

Table 4-3

5. Turn off the voltage source and switch the DMM from the voltage range to the current range. Disconnect the circuit and insert the ammeter in series with the circuit as shown in Figure 4-4.

Points to Note:	The ammeter must be placed in series with the circuit, allowing the circuit current to pass through the meter. It must never be connected across an element, since this will result in a *short circuit* and may damage the meter or the circuit.

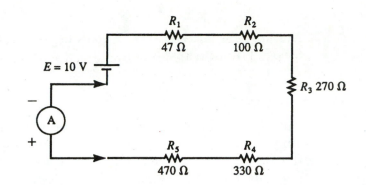

Figure 4-4 Current measurement

Turn the voltage source back on and record the circuit current in the space provided here. If all resistors have a 5% tolerance, you should measure a current which is within 5% of the calculated value of Table 4-2.

I	

6. Obtain a second voltage source or use the second supply of a dual power supply. Modify the series circuit by placing the second source in a *series-aiding connection* with the first voltage source as illustrated in Figure 4-5. Adjust the second supply for 5 V. (The first supply is still 10 V.) Notice that in the series-aiding connection, the positive terminal of the second source is connected to the negative terminal of the first source. Measure and record the voltage drop across each resistor.

	Voltage
V_1	
V_2	
V_3	
V_4	
V_5	

Table 4-4

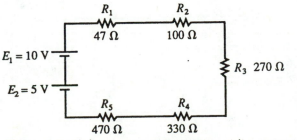

Figure 4-5 Voltage sources in a series-aiding connection

7. Measure the current in the circuit using the method described in Step 5. Enter the result here.

I	

8. Turn both voltage sources off and reverse the terminals of the 5-V supply as shown in Figure 4-6. Turn both supplies on and adjust the voltages if necessary. The voltage sources are now in a *series-opposing connection*. Examine the circuit of Figure 4-6 and determine the correct direction of current. Measure and record the voltage drop across each resistor.

	Voltage
V_1	
V_2	
V_3	
V_4	
V_5	

Table 4-5

9. Measure the current in the circuit using the method described in Step 5. Enter the result here.

I	

Ground

Electrical and electronic circuits are often connected to *ground,* meaning that this part of the circuit is at the same potential as the ground connection on a three-terminal plug. Since all grounds are connected in a building's electrical panel (as well as the water pipes), special precautions may need to be followed when using instru-

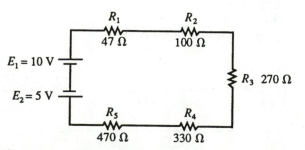

Figure 4-6   Voltage sources in a series-opposing connection

ments which are also grounded. Most dc voltage sources have a separate ground terminal at the front of the instrument, which permit us to connect a circuit to ground.

10. Refer to Figure 4-7(a). Connect the common terminals between the voltage sources to the ground terminal by using either a jumper wire or the grounding strip provided with the voltage source(s). Figure 4-7(b) indicates an alternate way of representing the voltage sources as *point sources*. Point sources simply indicate the potential at a point with respect to the reference (or in this case, ground). The ground symbol may or may not be shown. Connect the common (–) terminal of the DMM (voltmeter) to the ground point. All further voltage measurements are taken with respect to this point. Connect the voltage (+) terminal of the voltmeter to measure E_1, E_2, V_a, V_b, V_c, V_d. Record your results in Table 4-6.

	Voltage
E_1	
E_2	
V_a	
V_b	
V_c	
V_d	

Table 4-6

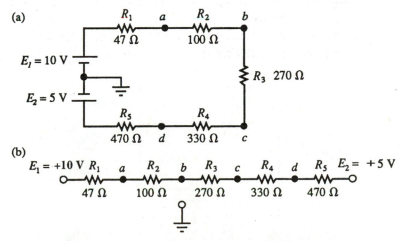

Figure 4-7 Ground connections and point sources

Internal Resistance of Voltage Sources

All voltage sources have some internal resistance, which tends to reduce the voltage between the terminals of the source when the source is under load. The internal resistance of the voltage sources used up to now have very small internal resistance (typically less than 1 Ω). Other sources such as nickel-cadmium and alkaline batteries have relatively large internal resistance. This means that as a circuit requires more current, the voltage across the terminals of the battery will decrease. The magnitude of the internal resistance determines the maximum amount of usable current that a voltage source can provide to a circuit.

11. Use a DMM to measure the voltage across the terminals of a 1.5-V D-cell alkaline battery and a 9-V alkaline battery. Enter the results here.

E(1.5-V cell)	
E(9-V cell)	

12. Using each of the resistors from the previous part of the lab, construct the circuit shown in Figure 4-8. Measure and record the voltage across the terminals of the batteries for each of the resistors in Table 4-7.

CAUTION:	When connecting some of the resistors across the 9-V battery, the 1/4-W power rating will be exceeded. You will need to use 2-W resistors to permit voltage measurements with the 47-Ω, 100-Ω and 270-Ω resistors. Since the resistors may get quite hot, extra precaution should be observed when handling them.

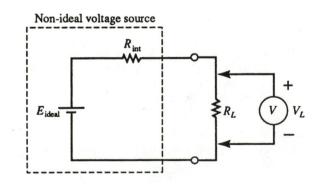

Figure 4-8 Measuring internal resistance of a voltage source

	V_L (1.5-V cell)	V_L (9-V cell)
$R_L = 470 \ \Omega$		
$R_L = 330 \ \Omega$		
$R_L = 270 \ \Omega$		
$R_L = 100 \ \Omega$		
$R_L = 47 \ \Omega$		

Table 4-7

CONCLUSIONS

13. Compare the measured resistance of Step 2 to the theoretical resistance recorded in Table 4-2. Determine the percentage variation as shown.

    R_T (theoretical) = _____ R_T (measured) = _____

    $$\text{percent variation} = \frac{\text{Measurement} - \text{Theoretical}}{\text{Theoretical}} \times 100\% \quad (4\text{-}4)$$

    percent variation = _____

14. Compare the measured voltage drops of Table 4-3 to the theoretical values recorded in Table 4-2. Indicate which values (if any) have a variation more than the resistor tolerance. Offer an explanation.

    _____

    _____

15. Examine the measured current of Step 5 and compare it to the theoretical value recorded in Table 4-2. Determine the percentage variation.

    percent variation = _____

16. Determine the summation of the voltage drops recorded in Table 4-4. Compare this value to the summation of voltage rises. Is Kirchhoff's voltage law satisfied?

    $\Sigma V =$ _____ $\Sigma E =$ _____

17. Calculate the theoretical current for the circuit of Figure 4-5. Compare this value to the measured value and determine the percent variation.

$I_{\text{theoretical}} =$ _____   percent variation = _____

18. Determine the summation of the voltage drops recorded in Table 4-5. Compare this value to the summation of voltage rises. Is Kirchhoff's voltage law satisfied?

$\Sigma V =$ _____ $\Sigma E =$ _____

_____

_____

19. Calculate the theoretical current for the circuit of Figure 4-6. Compare this value to the measured value and determine the percent variation.

$I_{\text{theoretical}} =$ _____   percent variation = _____

20. Calculate the voltages V_{ab}, V_{bc}, and V_{cd} from your measurements in Table 4-6. Compare your calculations to the corresponding measurements recorded in Table 4-5. Notice the direct correlation.

$V_{ab} = V_a - V_b =$ _____ $V_2 =$ _____

$V_{bc} =$ _____ $V_3 =$ _____

$V_{cd} =$ _____ $V_4 =$ _____

21. For each load resistor, evaluate the internal resistance of the 1.5-V battery by applying Ohm's law as shown below. Enter your results in Table 4-8. Use these results to determine the average value of internal resistance.

$$I = \frac{V_L}{R_L} \qquad (4\text{-}5)$$

$$R_{\text{int}} = \frac{E_{\text{ideal}} - V_L}{I} = \left(\frac{E_{\text{ideal}} - V_L}{V_L}\right) R_L \qquad (4\text{-}6)$$

R_L	R_{int}
470 Ω	
330 Ω	
270 Ω	
100 Ω	
47 Ω	
Average value of R_{int}	

Table 4-8

22. For each load resistor, use equations (4-5) and (4-6) to evaluate the internal resistance of the 9-V battery. Enter your results in Table 4-9. Use these results to determine the average value of internal resistance.

R_L	R_{int}
470 Ω	
330 Ω	
270 Ω	
100 Ω	
47 Ω	
Average value of R_{int}	

Table 4-9

Parallel dc Circuits

OBJECTIVES

After completing this lab, you will be able to
* assemble a parallel circuit consisting of a voltage source and several resistors,
* measure voltage and current in a parallel circuit,
* compare measured values to theoretical calculations and verify Kirchhoff's current law.
*

EQUIPMENT REQUIRED

☐ Digital multimeter (DMM)
☐ dc power supply
Note: Record this equipment in Table 5-1.

COMPONENTS

☐ Resistors: 470-Ω, 680-Ω, 1-kΩ, 2.2-kΩ, 4.7-kΩ (1/4-W, 5%)

EQUIPMENT USED

Instrument	Manufacturer/Model No.	Serial No.
DMM		
dc Supply		

Table 5-1

REFERENCE

Section 6.1   PARALLEL CIRCUITS
Section 6.2   KIRCHHOFF'S CURRENT LAW
Section 6.3   RESISTORS IN PARALLEL
Section 6.5   CURRENT DIVIDER RULE
Section 6.6   ANALYSIS OF PARALLEL CIRCUITS

DISCUSSION

Two elements are said to be in a *parallel* connection if they have exactly two nodes in common. Each element in a parallel circuit, as shown in Figure 5-1, has the same voltage across it.

The equivalent conductance of n resistors in parallel is determined as the summation of conductance

$$G_T = G_1 + G_2 + \dots + G_n \tag{5-1}$$

where the conductance G of each resistor is found as the reciprocal of resistance

$$G_x = \frac{1}{R_x} \tag{5-2}$$

The total resistance of n resistors in parallel is then found as

$$R_T = \frac{1}{G_T} \tag{5-3}$$

When a parallel network of resistors is connected in parallel with a voltage source, the current through the voltage source is determined as

$$I = \frac{E}{R_T} \tag{5-4}$$

The current through any resisistor in a parallel circuit is calculated using Ohm's law or the current divider rule, namely

$$I_x = \frac{E}{R_x} = \frac{R_T}{R_x} I \tag{5-5}$$

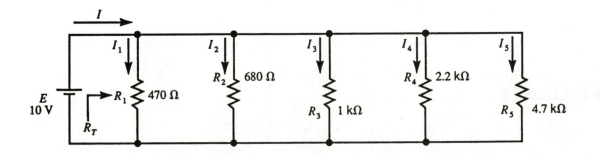

Figure 5-1   Parallel circuit

CALCULATIONS

1. Refer to the the the circuit of Figure 5-1. Calculate R_T, I, I_1, I_2, I_3, I_4, and I_5. Enter the results in Table 5-2. Show the correct units for each entry.

I_1	
I_2	
I_3	
I_4	
I_5	

R_T	
I	

Table 5-2

MEASUREMENTS

2. Connect the resistors as shown in the network of Figure 5-2. Use the DMM (ohmmeter) to measure the resistance across the open terminals. Enter the result here. Compare your measurement to the theoretical calculation in Table 5-2. You should observe only a small discrepancy.

R_T	

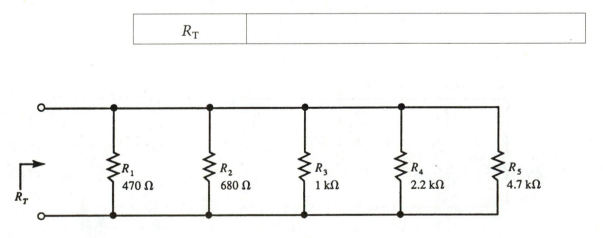

Figure 5-2   Parallel resistance

3. Connect the voltage source to the circuit as shown in Figure 5-3. With a DMM (voltmeter) connected across the voltage source, adjust the voltage for exactly 10 V.

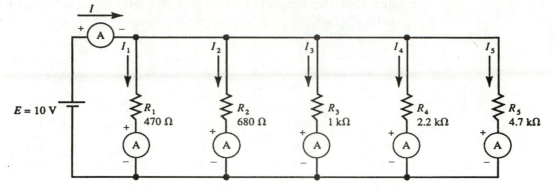

Figure 5-3   Measuring current in a parallel circuit

4. Disconnect the voltmeter from the voltage source. Turn the power supply off, without disturbing the voltage setting on the power supply. Set the DMM to measure current. Sucessively disconnect each branch of the circuit and insert the ammeter into the branch as illustrated in Figure 5-3. Measure the current through the voltage source and through each resistor in the circuit. **Ensure that each branch is reconnected after the ammeter is removed**. Record your results in Table 5-3.

	Current
I	
I_1	
I_2	
I_3	
I_4	
I_5	

Table 5-3

KIRCHHOFF'S CURRENT LAW

Kirchhoff's voltage and current laws provide an important foundation for the analysis of circuits. Kirchhoff's current law states:

The summation of currents entering a node is equal to the summation of currents leaving the node.

We now examine how Kirchhoff's current law can be verified in a laboratory.

5. Relocate the ammeter as shown in Figure 5-4 and measure the currents I_6, I_7, and I_8. Record your results in Table 5-4.

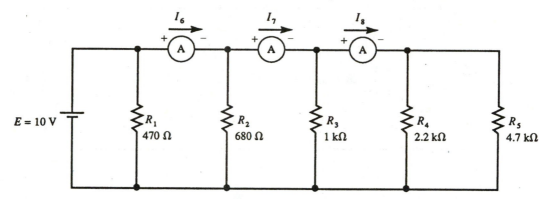

Figure 5-4   Verifying Kirchhoff's current law

	Current
I_6	
I_7	
I_8	

Table 5-4

CONCLUSIONS

6. Compare the measured resistance of Step 2 to the theoretical resistance recorded in Table 5-2. Determine the percentage variation as shown.

R_T (theoretical) = _____ R_T (measured) = _____

$$\text{percent variation} = \frac{\text{Measurement} - \text{Theoretical}}{\text{Theoretical}} \times 100\% \quad (5\text{-}6)$$

percent variation = _____

7. Compare the measured currents of Step 4 to the theoretical values recorded in Table 5-2. Indicate which values (if any) have a variation more than the resistor tolerance. Offer an explanation.

_____

_____

8. Refer to the data of Table 5-3 and Table 5-4.

   a. Compare current I to the summation $I_1 + I_2 + I_3 + I_4 + I_5$.

   $I =$ _____ $I_1 + I_2 + I_3 + I_4 + I_5 =$ _____

   b. Compare current I_6 to the the summation $I_2 + I_3 + I_4 + I_5$.

   $I_6 =$ _____ $I_2 + I_3 + I_4 + I_5 =$ _____

   c. Compare current I_7 to the the summation $I_3 + I_4 + I_5$.

   $I_7 =$ _____ $I_3 + I_4 + I_5 =$ _____

   d. Compare current I_8 to the the summation $I_4 + I_5$.

   $I_8 =$ _____ $I_4 + I_5 =$ _____

   e. Do the above calculations and measurements verify Kirchhoff's current law? Explain your answer.

   _____

   _____

LAB 6

Series-Parallel dc Circuits

OBJECTIVES

After completing this lab, you will be able to
- assemble a series-parallel circuit consisting of a voltage source and several resistors,
- measure voltage and current in a series-parallel circuit,
- compare measured values to theoretical calculations and verify Kirchhoff's current and voltage laws,
- assemble a zener diode regulator circuit and measure voltages and currents to verify that Kirchhoff's voltage and current laws apply,
- calculate power and verify the law of conservation of energy.

EQUIPMENT REQUIRED

☐ Digital multimeter (DMM)
☐ dc power supply
Note: Record this equipment in Table 6-1.

COMPONENTS

☐ Resistors: 330-Ω, 470-Ω, 680-Ω, 1-kΩ, 2.2-kΩ, 4.7-kΩ
                 (1/4-W, 5%)
☐ Zener diode: 1N4734A (1-W, 5.6-V$\pm$5%)

EQUIPMENT USED

Instrument	Manufacturer Model No.	Serial No.
DMM		
dc Supply		

Table 6-1

TEXT REFERENCE

Section 7.1   THE SERIES-PARALLEL NETWORK
Section 7.2   ANALYSIS OF SERIES-PARALLEL CIRCUITS
Section 7.3   APPLICATIONS OF SERIES-PARALLEL CIRCUITS

DISCUSSION

Regardless of the complexity of a circuit, the basic laws of circuit analysis always apply. While Ohm's law and Kirchhoff's voltage and current laws are used to analyze simple series and parallel circuits, these same laws may be applied to analyze even the most complicated circuit. The following rules apply to all circuits.

The same current occurs through all series elements.

The same voltage appears across all parallel elements.

CALCULATIONS

1. Refer to the the circuit of Figure 6-1. Calculate the total resistance, R_T, seen by the voltage source. Calculate the current, I. Solve for all resistor currents, voltages, and powers. Enter the results in Table 6-2. Show the correct units for all entries.

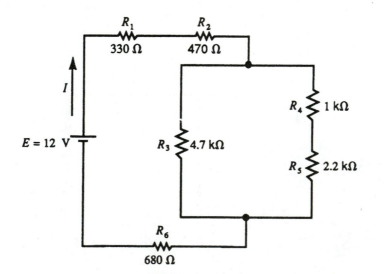

Figure 6-1   Series-parallel circuit

	Current	Voltage	Power
R_1			
R_2			
R_3			
R_4			
R_5			
R_6			

R_T	
I	

Table 6-2

MEASUREMENTS

2. Connect the resistors as shown in the network of Figure 6-2. Use the DMM (ohmmeter) to measure the resistance across the open terminals. Enter the result here. Compare your measurement to the theoretical calculation in Table 6-2. You should observe only a small discrepancy.

R_T	

3. Connect the voltage source to the circuit as shown in Figure 6-1. With a DMM (voltmeter) connected across the voltage source, adjust the voltage for exactly 12 V.

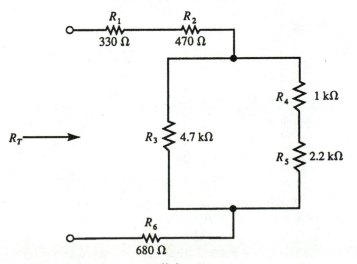

Figure 6-2 Series-parallel resistance

4. Disconnect the voltmeter from the voltage source. Measure the voltage across each resistor in the circuit and record the results in Table 6-3.
5. Set the DMM to measure the currents as illustrated in Figure 6-3. **Ensure that each branch is reconnected after the ammeter is removed.** Record your measurements in Table 6-3.

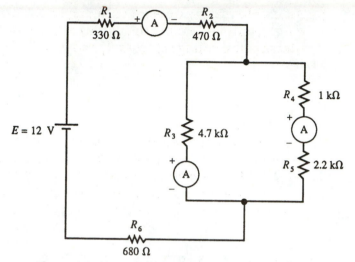

Figure 6-3   Measuring current in a series-parallel circuit

	Current	Voltage
R_1		
R_2		
R_3		
R_4		
R_5		
R_6		

I	

Table 6-3

Zener Diode Circuit

In this part of the lab, we apply the principles of circuit analysis to examine the operation of a more complicated circuit. Here we use a *zener diode*, which is a two-terminal semiconductor device normally used as a *voltage regulator* to maintain a constant voltage between two terminals. When the zener diode is placed across a component which has a voltage greater than the *break-over* (zener) voltage of the diode, current through the zener diode forces the voltage across the component to decrease. While most other diodes permit current in only the forward direction (in the direction of the arrow in the diode symbol), the zener diode can conduct in either direction. When used as a voltage regulator, the zener diode is operated in the reverse-biased conditon. This means that when the zener diode is in its *breakover region*, current is against the arrow of the diode symbol.

6. Assemble the circuit shown in Figure 6-4, temporarily omitting the zener diode. Adjust the voltage source for 12 V.
7. Measure the voltages across R_1 and R_2 and record the values here.

V_1	
V_2	

8. Insert the zener diode into the circuit. Use the DMM (voltmeter) to measure voltages V_1 and V_2 in the circuit of Figure 6-4. Convert the DMM to measure current and correctly measure currents I_1, I_2, and I_Z. **Make sure that you turn off the voltage supply before disconnecting the circuit to insert the ammeter.** In the circuit of Figure 6-4, show where you placed the ammeters to measure the currents. Record all measurements in Table 6-4.

Current	Voltage
$I_1 =$	$V_1 =$
$I_2 =$	$V_2 =$
$I_Z =$	

Table 6-4

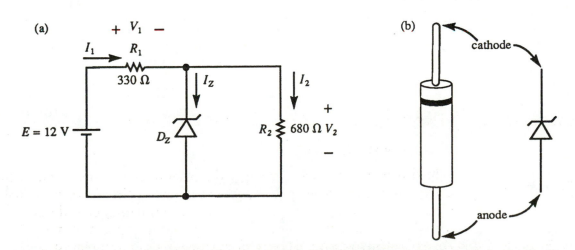

Figure 6-4   Zener diode voltage regulator circuit

CONCLUSIONS

9. Compare the measured resistance of Step 2 to the theoretical value recorded in Table 6-2. Determine the percent variation.

R_T (theoretical) = _____ R_T (measured) = _____

percent variation = _____

10. Compare the measured voltage drops in Table 6-3 to the theoretical values recorded in Table 6-2. Indicate which values (if any) have a variation more than the resistor tolerance. Offer an explanation.

_____

_____

11. Compare the measured currents in Table 6-3 to the theoretical values recorded in Table 6-2. Indicate which values (if any) have a variation more than the resistor tolerance. Offer an explanation.

_____

_____

12. Refer to the data of Table 6-3.

    a. Compare current I to the the summation $I_3 + I_4$

    $I =$ _____ $I_3 + I_4 =$ _____

    b. Do the above calculations and measurements satisfy Kirchhoff's current law? Explain your answer.

13. a. Use the data of Table 6-3 to calculate the power, P_T, delivered to the circuit by the voltage source. Enter the result here.

P_T	

   b. Use the voltages and currents of Table 6-3 to determine the power dissipated by each resistor in the circuit. Enter your results in Table 6-5.

	Power
R_1	
R_2	
R_3	
R_4	
R_5	
R_6	

Table 6-5

    c. Compare the total power dissipated by the resistors to the total power delivered by the voltage source. Is energy conserved?

_____

_____

14. Refer to the circuit of Figure 6-4. Compare the voltage, V_2, with the zener diode removed from the circuit, to the voltage when the diode is in the circuit. How do they compare? Explain briefly why this occurred.

_____

_____

15. Refer to the data of Table 6-4.
    a. Compare current I_1 to the the summation $I_2 + I_Z$.

$I_1 = $ _____ $I_2 + I_Z = $ _____

    b. Do the calculations and measurements in part a satisfy Kirchhoff's current law? Explain your answer.

_____

_____

16. a. Use the data in Table 6-4 to calculate the total power, P_T, de-

livered to the circuit by the voltage source. Enter the result here.

P_T	

b. Use the voltages and currents of Table 6-4 to determine the power dissipated by the zener diode and by each resistor in the circuit. Enter your results in Table 6-6.

	Power
R_1	
R_2	
D_Z	

Table 6-6

c. Compare the total power dissipated by the resistors to the total power delivered by the voltage source. Is energy conserved?

_____

_____

PROBLEMS

17. Refer to the circuit of Figure 6-4. Assume that the zener diode has a break-over voltage of 4.3 V.
    a. Calculate the voltages V_1 and V_2
    b. Determine the currents I_1, I_2 and I_Z
    c. Solve for the powers dissipated by R_1, R_2, and D_Z.
    d. Show that the total power dissipated is equal to the power delivered by the voltage source.

Voltmeter Loading Effects

OBJECTIVES

After completing this lab, you will be able to
- use a voltmeter's loading resistance to predict the voltage that will be indicated when measuring a high-resistance circuit,
- use the sensitivity S of an analog meter to determine the loading resistance of a voltmeter,
- measure and compare the loading effects of analog and digital voltmeters in high-resistance circuits,
- measure the loading effects of voltmeters in low-resistance circuits.

EQUIPMENT REQUIRED

- ☐ Digital multimeter (DMM)
- ☐ Analog multimeter (VOM)
- ☐ dc power supply
  Note: Record this equipment in Table 7-1.

COMPONENTS

- ☐ Resistors: 3.3-kΩ, 5.6-kΩ, 3.3-MΩ, 5.6-MΩ (1/4-W carbon, 5%)

EQUIPMENT USED

Instrument	Manufacturer/Model No.	Serial No.
DMM		
VOM		
dc Supply		

Table 7-1

TEXT REFERENCE

Section 6.8   VOLTMETER LOADING EFFECTS
Section 7.5   LOADING EFFECTS OF INSTRUMENTS

DISCUSSION

When we use a voltmeter to measure the voltage in a circuit, it is natural to assume that if the meter is operating correctly, it will always indicate the correct voltage in a circuit. Unfortunately this is not always the case.

In order for any instrument to provide a measurement, it must take a small amount of energy from the circuit under test, and use this energy to provide a reading. While the amount of energy taken from most circuits is virtually undetectable, this is not always so. Examine the circuit of Figure 7-1. Ideally, the internal resistance of a voltmeter is infinitely large, resulting in no circuit current. The voltage appearing across the voltmeter will be 10 V which means that the voltmeter provides a correct reading of 10.0 V.

However, all voltmeters have some internal resistance. If the internal resistance of the voltmeter were equal to the series resistance, R_1, then the voltage appearing across the voltmeter would be half of the supply voltage, resulting in a reading of 5.0 V. If, on the other hand, the internal resistance of the meter was $R_1/4$, then the voltage across the voltmeter would only be 2.0 V. The degree to which a meter loads a circuit under test is called the *loading effect* and is determined mathematically as follows:

$$\text{loading effect} = \frac{\text{unloaded value} - \text{reading}}{\text{unloaded value}} \times 100\% \qquad (7\text{-}1)$$

CALCULATIONS

1. The internal resistance of most digital voltmeters is typically several megohms, while the resistance for most analog voltmeters is somewhat lower and generally depends on the voltage range. Refer to the circuit of Figure 7-1. Determine the expected voltage reading and the loading effect (in percent) for a DMM which has an internal resistance of 10 MΩ. Repeat the calulations for a VOM which has an internal resistance of 5 MΩ. Record your results in Table 7-2.

	Reading	Loading effect
DMM (10 MΩ)		
VOM (5 MΩ)		

Table 7-2

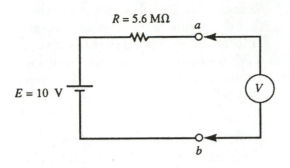

Figure 7-1   Voltmeter loading effect of a high-resistance source

MEASUREMENTS

2. Refer to the manufacturer's specifications for your DMM. Obtain the internal resistance of the voltmeter and record this value.

R_{int}	

3. The internal resistance of an analog meter is generally dependent on the voltage range used. In order to determine the internal resistance, the manufacturer provides a specification called the *sensitivity*, *S*, which has units of kΩ/V. The internal resistance is then determined as the product of sensitivity and the voltage range of the meter as follows:

$$R_{int} = SV_{range} \tag{7-2}$$

Refer to the manufacturer's specifications for your VOM and obtain the sensitivity of the meter. Determine the correct range that you would use to measure a voltage of 10 V. Calculate and record the internal resistance of your meter on this range.

S	
R_{int}	

4. Assemble the circuit of Figure 7-1. Place the DMM voltmeter directly across the terminals of the voltage source and adjust the voltage source for exactly 10 V. Remove the voltmeter from the voltage source. Since the circuit has no load connected between terminals *a* and *b*, the unloaded voltage between these terminals must be 10 V. Now connect the DMM between terminals *a* and *b*. Measure the voltage appearing between these terminals. You should notice a measurable difference. Record the results.

V_{ab} (unloaded)	
V_{ab} (measured)	

5. Remove the DMM from the circuit of Figure 7-1. Place the VOM on the appropriate range to measure 10 V. Use the DMM (ohm-meter) to measure the resistance between the terminals of the VOM voltmeter. Record the result here. The measured resistance should be equal to the internal resistance calculated in Step 3.

R_{int}	

6. Insert the VOM between terminals a and b. Measure the voltage appearing between these terminals. Record the results.

V_{ab} (unloaded)	
V_{ab} (measured)	

7. Replace the 5.6-MΩ resistor in the circuit of Figure 7-1 with a 5.6-kΩ resistor and repeat Step 4. (You should observe only minimal loading effect.)

V_{ab} (unloaded)	
V_{ab} (measured)	

8. Repeat Step 6 using the 5.6-kΩ resistor.

V_{ab} (unloaded)	
V_{ab} (measured)	

9. Refer to the circuit of Figure 7-2. Use the voltage divider rule to calculate the voltage across each of the resistors. Record the results here.

V_1 (unloaded)	
V_2 (measured)	

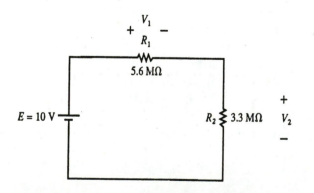

Figure 7-2   Voltmeter loading effect of a high-resistance circuit

10. Assemble the circuit of Figure 7-2 and measure the voltages V_1 and V_2 with both the DMM and VOM voltmeters. Indicate which range of the VOM you used. (You will observe a large loading effect with both meters).

	DMM	VOM Range:
V_1 (measured)		
V_2 (measured)		

11. Refer to the circuit of Figure 7-3. Use the voltage divider rule to calculate the voltage across each of the resistors. Record the results here.

V_1	
V_2	

12. Assemble the circuit of Figure 7-3 and measure the voltages V_1 and V_2 with both the DMM and VOM voltmeters. Indicate which range of the VOM you used. (You should observe minimal loading effect with both meters.)

	DMM	VOM Range:
V_1 (measured)		
V_2 (measured)		

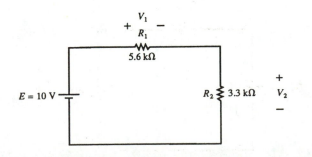

Figure 7-3 Voltmeter loading effect of a low-resistance circuit

CONCLUSIONS

13. Using the manufacturer's specified internal resistance of the DMM, determine the voltage which should appear between terminals a and b when the voltmeter is placed into the circuit of Figure 7-1. Calculate the theoretical loading effect for the meter.

    V_{ab} (loaded) = _____ Loading effect = _____

14. Use the measurements of Step 4 to calculate the actual loading effect of the meter. Compare the result to the theoretical value of Step 13.

    Loading effect = _____

    _____

    _____

15. Using the actual internal resistance of the VOM, determine the voltage which should appear between terminals a and b when the voltmeter is placed into the circuit of Figure 7-1. Calculate the theoretical loading effect for the meter.

    V_{ab} (loaded) = _____   Loading effect = _____

16. Use the measurements of Step 6 to calculate the actual loading effect of the meter. Compare the result to the theoretical value of Step 15.

    Loading effect = _____

    _____

17. Use the measurements of Step 7 to calculate the actual loading effect of the DMM when the source resistance is 5.6 kΩ.

Loading effect = _____

18. Use the measurements of Step 8 to calculate the actual loading effect of the VOM when the source resistance is 5.6 kΩ.

Loading effect = _____

19. Using the manufacturer's specified internal resistance of the DMM, determine the voltage which should appear across each of the resistors when the voltmeter is placed into the circuit of Figure 7-2. Calculate the theoretical loading effect for each reading.

V_1 (loaded) = _____ V_1 Loading effect = _____

V_2 (loaded) = _____ V_2 Loading effect = _____

20. Use the measurements of Step 10 to calculate the actual loading effect of the DMM voltmeter. Compare the results to the theoretical value of Step 19.

V_1 Loading effect = _____

V_2 Loading effect = _____

---

21. Using the sensitivity of the VOM and the voltage range of Step 10, calculate the internal resistance of the VOM. Determine the voltage which should appear across each of the resistors when the VOM voltmeter is placed into the circuit of Figure 7-2. Calculate the theoretical loading effect for each reading.

R_{int} = _____

V_1 (loaded) = _____ V_1 Loading effect = _____

V_2 (loaded) = _____ V_2 Loading effect = _____

22. Use the measurements of Step 10 to calculate the actual loading effect of the VOM voltmeter. Compare the results to the theoretical value of Step 21.

V_1 Loading effect = _____

V_2 Loading effect = _____

23. Use the measurements of Step 12 to calculate the loading effect of the DMM voltmeter in the circuit of Figure 7-3.

V_1 Loading effect = _____

V_2 Loading effect = _____

24. Use the measurements of Step 12 to calculate the loading effect of the VOM voltmeter in the circuit of Figure 7-3.

V_1 Loading effect = _____

V_2 Loading effect = _____

25. For high-resistance circuits, which voltmeter (the DMM or the VOM) has the higher loading effect?

_____

26. How does the loading effect of a voltmeter in a high-resistance circuit compare to the loading effect in a low-resistance circuit?

_____

_____

Potentiometers and Rheostats

OBJECTIVES

After completing this lab, you will be able to
- demonstrate the use a variable resistor as a potentiometer to control the voltage applied to a load,
- demonstrate the use a variable resistor as a rheostat to control the current applied to a load,
- measure how the value of a load's resistance affects the voltage across a potentiometer.

EQUIPMENT REQUIRED

☐ Digital multimeter (DMM)
☐ dc power supply
   Note: Record this equipment in Table 8-1.

COMPONENTS

☐ Resistors: 5.6-kΩ, 3.3-kΩ, 330-kΩ (1/4-W, 5%)
                 10-kΩ variable resistor

EQUIPMENT USED

Instrument	Manufacturer/Model No.	Serial No.
DMM		
dc Supply		

Table 8-1

TEXT REFERENCE

Section 3.5   TYPES OF RESISTORS
Section 7.4   POTENTIOMETERS

63

DISCUSSION

Variable resistors are used extensively in electrical and electronic circuits to control the voltage and current in circuits. When a variable resistor is used to control the voltage (as in the volume control of an amplifier) it is called a *potentiometer*. If the same resistor is used to control the amount of current through a circuit (such as in a light dimmer) it is called a *rheostat*.

Refer to the series circuit of Figure 8-1. The current through this circuit is constant regardless of the location of the wiper arm (terminal *b*) of the variable resistor. If the wiper arm is moved so that it is at the bottom of the resistor, the resistance between terminals *b* and *c* is zero. This results in the voltage, V_L, being zero volts. If, however, the wiper arm is moved so that it is at the top of the resistor, the resistance between terminals *b* and *c* will be at a maximum, resulting in a maximum voltage, V_L, appearing between the terminals.

CALCULATIONS

1. Determine the range of the output voltage V_L for the circuit of Figure 8-1. Calculate the range of output voltage if a 330-kΩ resistor is connected across the output terminals of the circuit. Recalculate the range of output voltage if a 3.3-kΩ resistor is connected across the output terminals of the circuit. Record your results in Table 8-2.

	V_L (min)	V_L (max)
$R_L = \infty$ (open)		
$R_L = 330$ kΩ		
$R_L = 3.3$ kΩ		

Table 8-2

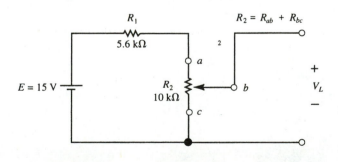

Figure 8-1   Variable resistor used as a potentiometer

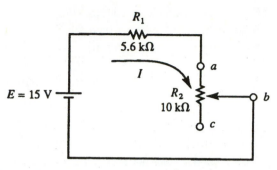

Figure 8-2 Variable resistor used as a rheostat

2. The 10-kΩ resistor in the circuit of Figure 8-1 is easily converted from a potentiometer into a rheostat. Figure 8-2 shows how the variable resistor is used as a rheostat. When the wiper arm is moved so that it is at the top of the resistor, the resistance of the rheostat will be at its minimum value. This means that maximum current will occur in the circuit. If the wiper is moved to the bottom of the resistor, the resistance of the rheostat will be at its maximum value, resulting in the least amount of current.

   Calculate the range of current for the circuit of Figure 8-2.

I(min)	I(max)

MEASUREMENTS

3. Assemble the circuit shown in Figure 8-1. Adjust the voltage source for 15 V. With a voltmeter connected between terminals b and c of the potentiometer, use a small screwdriver to adjust the central wiper fully clockwise (CW). Measure and record the output voltage. Now adjust the central wiper fully counterclockwise (CCW). Again measure and record the output voltage.

V_L(CW)	V_L(CCW)

4. Adjust the potentiometer to obtain an output voltage of 3.0 V. Disconnect the potentiometer from the circuit, being careful not to readjust the potentiometer setting. Measure and record the resistance between terminals b and c of the potentiometer.

R_{bc}	

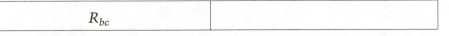

5. Connect a 330-kΩ load resistor between the output terminals. Adjust the voltage source for 15 V. With a voltmeter connected between terminals *b* and *c* of the potentiometer, use a small screwdriver to adjust the central wiper fully clockwise (CW). Measure and record the output voltage. Now adjust the central wiper fully counterclockwise (CCW). Again measure and record the output voltage.

V_L(CW)	V_L(CCW)

6. Adjust the potentiometer to obtain an output voltage of 3.0 V. Disconnect the potentiometer from the circuit, being careful not to readjust the potentiometer setting. Measure and record the resistance between terminals *b* and *c* of the potentiometer.

R_{bc}	

7. Connect a 3.3-kΩ load resistor between the output terminals. Adjust the voltage source for 15 V. With a voltmeter connected between terminals *b* and *c* of the potentiometer, use a small screwdriver to adjust the central wiper fully clockwise (CW). Measure and record the output voltage. Now adjust the central wiper fully counterclockwise (CCW). Again measure and record the output voltage.

V_L(CW)	V_L(CCW

8. Adjust the potentiometer to obtain an output voltage of 3.0 V. Disconnect the potentiometer from the circuit, being careful not to readjust the potentiometer setting. Measure and record the resistance between terminals *b* and *c* of the potentiometer.

R_{bc}	

9. Construct the circuit of Figure 8-2. Place a DMM ammeter into the circuit to measure the current. Use a small screwdriver to adjust the central wiper fully clockwise (CW). Measure and record the circuit current. Now adjust the central wiper fully counterclockwise (CCW). Again measure and record the current.

I(CW)	I(CCW)

10. Adjust the rheostat so that the measured current is exactly equal to $I_{max}/2$. Remove the rheostat from the circuit, being careful not to readjust the potentiometer setting. Measure and record the resistance between terminals a and b of the potentiometer. The resistance should be exactly equal to R_1.

R_{ab}	

CONCLUSIONS

11. Compare the measurements of Step 3 to the theoretical maximum and minimum values of voltage recorded in Table 8-2 for $R_L = \infty$ (open).

_____

_____

_____

_____

12. Compare the measurements of Step 5 to the theoretical maximum and minimum values of voltage recorded in Table 8-2 for $R_L = 330$ kΩ.

_____

_____

_____

_____

13. Use the measured resistance R_{bc} of Step 6 to determine the theoretical voltage which would appear across the load $R_L = 330$ kΩ in the equivalent circuit of Figure 8-3.

V_L	

Compare the above value to the measured load voltage V_L in Step 6.

_____

14. Compare the measurements of Step 7 to the theoretical maximum and minimum values of voltage recorded in Table 8-2 for $R_L = 3.3 \text{ k}\Omega$.

_____

_____

15. Use the measured resistance R_{bc} of Step 8 to determine the theoretical voltage which would appear across the load $R_L = 3.3 \text{ k}\Omega$ in the equivalent circuit of Figure 8-3.

V_L	

Compare the above value to the measured load voltage V_L in Step 8.

_____

16. Compare the measurements of Step 9 to the theoretical maximum and minimum values of current recorded in Step 2.

_____

_____

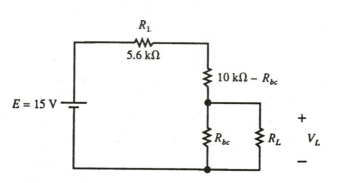

Figure 8-3

Ammeter Loading Effects

OBJECTIVES

After completing this lab, you will be able to
- calculate an ammeter's loading resistance and predict the current which will be indicated when measuring a low-resistance circuit,
- measure and compare the loading effects of analog and digital ammeters in low-resistance circuits,
- measure the loading effects of ammeters in high-resistance circuits,
- "measure" current in a low-resistance circuit without needing to consider loading effects of ammeters.

EQUIPMENT REQUIRED

☐ Digital multimeter (DMM)
☐ Volt-ohm-milliameter (VOM)
☐ dc power supply
   Note: Record this equipment in Table 9-1.

COMPONENTS

☐ Resistors: 4.7-Ω, 6.8-Ω, 10-Ω, 470-Ω, 680-Ω, 1-kΩ
                  (1/4-W, 5%)

EQUIPMENT USED

Instrument	Manufacturer/Model No.	Serial No.
DMM		
VOM		
dc Supply		

Table 9-1

TEXT REFERENCE

Section 5.12   AMMETER LOADING EFFECT
Section 7.5   LOADING EFFECTS OF INSTRUMENTS

DISCUSSION

In Lab #7, we examined the loading effects of voltmeters in circuits. It was observed that when a voltmeter is used to measure voltage in a high-resistance circuit, the internal resistance of the voltmeter affects the circuit. This *loading effect* results in a measured voltage which is less than the voltage would have been prior to connecting the voltmeter. A similar effect occurs when an ammeter is connected into a circuit.

Examine the circuit of Figure 9-1. When the ammeter is placed into the circuit to measure the current I, we expect that the ammeter will indicate the correct current. This would be a correct assumption if the ammeter had an internal resistance of zero ohms. (Ideally the ammeter would be a short circuit.) However, since all ammeters have some internal resistance there will always be some loading effect on the circuit. The amount of loading is dependent upon the resistance of the meter and on the equivalent circuit resistance. If the ammeter has an internal resistance which is at least ten times smaller than the circuit resistance, then its loading effect will be relatively small. Conversely, if the resistance of the meter is large in comparison to the circuit resistance, the loading effect will be large, resulting in a meaningless measurement.

Recall that the loading effect for any meter is given as follows:

$$\text{loading effect} = \frac{\text{unloaded value} - \text{reading}}{\text{unloaded value}} \times 100\% \qquad (9\text{-}1)$$

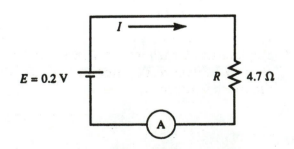

Figure 9-1 Measuring current in a low-resistance circuit

CALCULATIONS

1. The internal resistance of most digital ammeters is typically between 0.25 Ω and 5 Ω, depending on the range of the ammeter.
   Refer to the circuit of Figure 9-1. Determine the current reading
   and loading effect of a DMM ammeter which has an internal resistance of 5 Ω on its 20-mA range. Calculate the current reading and
   loading effect of a VOM ammeter which has an internal resistance of 2.5 Ω on its 100-mA range. Enter your results in Table 9-2.

	Reading	Loading effect
DMM (5 Ω)		
VOM (2.5 Ω)		

Table 9-2

MEASUREMENTS

2. Use a DMM ohmmeter to measure the actual resistance of a 4.7-Ω
   resistor. Assemble the circuit of Figure 9-1 and use a DMM voltmeter to adjust the voltage source for 0.200 V. Calculate the unloaded
   current in the circuit. Record the values here.

R(4.7-Ω)	
I(unloaded)	

3. Insert a DMM ammeter into the circuit. **Ensure that the ammeter is
   on the correct range to measure the current and that the meter is
   placed in series with the circuit**. Measure and record the current I.

I(DMM)	

4. Replace the DMM with a VOM ammeter. **Ensure that the ammeter is on the correct range to measure the current**. Measure and
   record the current I.

I(VOM)	

5. Replace the 4.7-Ω resistor of Figure 9-1 with a 10-Ω resistor and

repeat the measurements and calculations of Steps 2 to 4. Enter your results in Table 9-3.

$R(10\text{-}\Omega)$	
I(unloaded)	
I(DMM)	
I(VOM)	

Table 9-3

6. Obtain a 6.8-Ω resistor and measure its actual resistance value using a DMM ohmmeter. Assemble the circuit of Figure 9-2. Measure the voltage across each resistor. Since the DMM resistance is much greater than the circuit resistance, meter loading will be negligible. Record the results in Table 9-4.

$R(6.8\text{-}\Omega)$	
V_1	
V_2	
V_3	

Table 9-4

7. Use the measured results for voltage and resistance to calculate the unloaded currents I_1, I_2, and I_3. Record the results in Table 9-5.

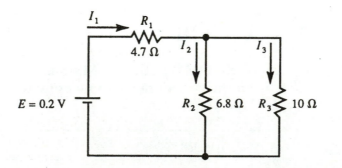

Figure 9-2 Current measurement in a low-resistance series-parallel circuit

I_1(unloaded)	
I_2(unloaded)	
I_3(unloaded)	

Table 9-5

8. Using the results from Table 9-5, set the DMM ammeter to the correct range and measure each of the currents in the circuit of Figure 9-2. You will notice some discrepancy due to meter loading. Record the results here.

I_1(reading)	
I_2(reading)	
I_3(reading)	

9. Repeat Step 8 using the VOM ammeter.

I_1(reading)	
I_2(reading)	
I_3(reading)	

10. Obtain a 470-Ω, a 680-Ω, and a 1-kΩ resistor. Measure and record each value using a DMM ohmmeter.

R(470-Ω)	
R(680-Ω)	
R(1-kΩ)	

11. Assemble the circuit of Figure 9-3 (next page). Measure the voltage across each resistor. Record the results in Table 9-6.

V_1	
V_2	
V_3	

Table 9-6

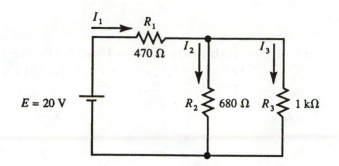

Figure 9-3   Current measurement in a high-resistance series-parallel circuit

12. Use the measured results for voltage and resistance to calculate the unloaded currents I_1, I_2, and I_3. Record your values in Table 9-7.

I_1(unloaded)	
I_2(unloaded)	
I_3(unloaded)	

Table 9-7

13. Using the results from Table 9-7, set the DMM ammeter to the correct range. Measure and record each of the currents in the circuit of Figure 9-3.

I_1(reading)	
I_2(reading)	
I_3(reading)	

14. Repeat Step 13 using the VOM ammeter.

I_1(reading)	
I_2(reading)	
I_3(reading)	

CONCLUSIONS

15. Refer to the data of Steps 2 and 3. Use equation (9-1) to calculate the loading effects for both the DMM ammeter and the VOM ammeter. Record them below.

DMM (loading)	
VOM (loading)	

16. Refer to the data of Step 5. Calculate the loading effects for both the DMM ammeter and the VOM ammeter. Record them below.

DMM (loading)	
VOM (loading)	

17. Based on the data of Steps 15 and 16, for which load resistor (the 4.7-Ω or the 10-Ω) was the loading effect more pronounced? Why do you think this occurred?

_____

_____

18. Refer to the data of Steps 7 through 9. Calculate the loading effects for each meter when used to measure the currents I_1, I_2, and I_3. Record the results in Table 9-8.

	DMM (loading)	VOM (loading)
I_1		
I_2		
I_3		

Table 9-8

19. Refer to the data of Steps 12 through 14. Calculate the loading effects for each meter when used to measure the currents I_1, I_2, and I_3. Record the results in Table 9-9.

	DMM (loading)	VOM (loading)
I_1		
I_2		
I_3		

Table 9-9

20. Although the circuits of Figures 9-2 and 9-3 have the same unloaded currents, the measured currents were not the same. In which circuit (Figure 9-2 or Figure 9-3) did the ammeters provide more accurate readings? What conclusion can you make about the loading effect of an ammeter as it relates to the resistance in the branch under test?

_____

_____

21. In Steps 7 and 12 you were able to determine the unloaded current in a circuit. Explain how it is possible to "measure" current in a low-resistance circuit without introducing loading error.

_____

_____

_____

22. The method used in Steps 7 and 12 has another advantage when compared to measuring current with an ammeter. Explain the advantage of "measuring" current by using a voltmeter across a known resistance.

_____

_____

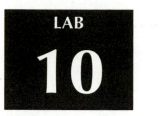

Superposition Theorem

OBJECTIVES

After completing this lab, you will be able to

* calculate currents and voltages in a dc circuit using the superposition theorem,
* measure voltage and current in a multi-source circuit,
* measure the effects of sucessively removing each voltage source from a circuit,
* calculate loop currents and node voltages using mesh analysis and nodal analysis,
* verify the superposition theorem as it applies to dc circuits and show that the results are consistent with results determined using mesh analysis and nodal analysis.

EQUIPMENT REQUIRED

☐ Digital multimeter (DMM)
☐ dc power supply (2)
Note: Record this equipment in Table 10-1.

COMPONENTS

☐ Resistors: 680-Ω, 1-kΩ, 3.3-kΩ (1/4-W, 5%)

EQUIPMENT USED

Instrument	Manufacturer/Model No.	Serial No.
DMM		
dc Supply		
dc Supply		

Table 10-1

TEXT REFERENCE

DISCUSSION

Mesh analysis allows us to find the loop currents for a circuit having any number of voltage or current sources. If a circuit contains current sources, these must first be converted to voltage sources.

Nodal analysis is the duality of mesh analysis in that it allows us to calculate nodal voltages of a circuit (with respect to a reference node). If a circuit contains voltage sources, it is necessary to first convert these to current sources.

Analyzing a circuit using mesh or nodal analysis usually requires solving several linear equations. The superposition theorem allows us to simplify the analysis of a multi-source circuit by considering only one source at a time.

The superposition theorem states:

The voltage across (or the current through) a resistor may be determined by finding the sum of the effects due to each independent source in the circuit.

In order to determine the effects due to one source, it is necessary to remove all other sources from the circuit. This is accomplished by replacing voltage sources with short circuits and by replacing current sources with open circuits.

CALCULATIONS

1. Use superposition to calculate currents I_1, I_2, and I_3 in the circuit of Figure 10-1 (next page). Record the results (next page).

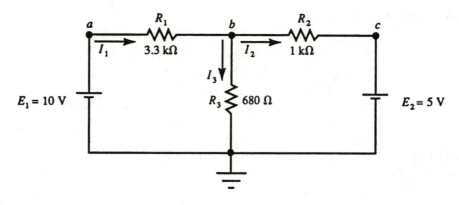

Figure 10-1

I_1	
I_2	
I_3	

2. Use superposition to calculate the currents I_1, I_2, and I_3 in the circuit of Figure 10-2. Record the results below.

I_1	
I_2	
I_3	

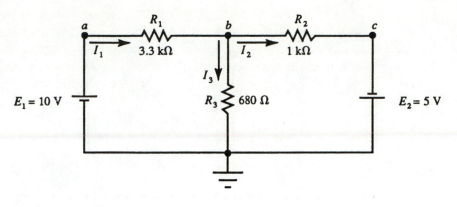

Figure 10-2

MEASUREMENTS

3. Assemble the circuit of Figure 10-1. Measure the voltages V_{ab}, V_b, and V_{bc}. Record your results in Table 10-2, showing the correct polarity for each measurement. Use these measurements and the resistor color codes to calculate the currents I_1, I_2, and I_3 for the circuit. Show the correct polarity for each current. (Use a negative sign to indicate that current is opposite to the indicated direction.)

V_{ab}	
V_b	
V_{bc}	
I_1	
I_2	
I_3	

Table 10-2

4. Assemble the circuit of Figure 10-2. Measure the voltages V_{ab}, V_b, and V_{bc}. Record your results in Table 10-3, showing the correct polarity for each measurement. Use these measurements and the resistor color codes to calculate the currents I_1, I_2, and I_3 for the circuit. Show the correct polarity for each current.

V_{ab}	
V_b	
V_{bc}	
I_1	
I_2	
I_3	

Table 10-3

5. Remove the voltage source E_2 from the circuit of Figure 10-1 and replace it with a short circuit. Measure the voltages V_{ab}, V_b, and V_{bc} and calculate the currents I_1, I_2, and I_3 due to the voltage source E_1. Record these results in Table 10-4.

$V_{ab(1)}$	
$V_{b(1)}$	
$V_{bc(1)}$	
$I_{1(1)}$	
$I_{2(1)}$	
$I_{3(1)}$	

Table 10-4

6. Remove voltage source E_1 from the circuit of Figure 10-1 and replace it with a short circuit. Measure voltages V_{ab}, V_b, and V_{bc} and use the results to calculate currents I_1, I_2, and I_3 due to the voltage source E_2. Record the results in Table 10-5.

$V_{ab(2)}$	
$V_{b(2)}$	
$V_{bc(2)}$	
$I_{1(2)}$	
$I_{2(2)}$	
$I_{3(2)}$	

Table 10-5

7. Reverse the polarity of voltage source E_2 as shown in the circuit of Figure 10-2. (Voltage source E_1 is still substituted by a short circuit.) Measure voltages V_{ab}, V_b, and V_{bc} and use the results to calculate currents I_1, I_2, and I_3 due to the voltage source E_2. Record the results in Table 10-6.

$V_{ab(3)}$	
$V_{b(3)}$	
$V_{bc(3)}$	
$I_{1(3)}$	
$I_{2(3)}$	
$I_{3(3)}$	

Table 10-6

CONCLUSIONS

8. Compare the measured currents of Table 10-2 to the theoretical currents calculated in Step 1. Determine the percent deviation for each of the currents.

I_1(theoretical) = _____   percent variation = _____

I_2(theoretical) = _____   percent variation = _____

I_3(theoretical) = _____   percent variation = _____

9. Compare the measured currents recorded in Table 10-3 to the theoretical currents calculated in Step 2. Determine the percent deviation for each of the currents.

I_1(theoretical) = _____   percent variation = _____

I_2(theoretical) = _____   percent variation = _____

I_3(theoretical) = _____   percent variation = _____

10. Combine the results of Table 10-4 and Table 10-5.

$$V_{ab} = V_{ab(1)} + V_{ab(2)} = \text{_____}$$

$$V_b = V_{b(1)} + V_{b(2)} = \text{_____}$$

$$V_{bc} = V_{bc(1)} + V_{bc(2)} = \text{_____}$$

$$I_1 = I_{1(1)} + I_{1(2)} = \text{_____}$$

$$I_2 = I_{2(1)} + I_{2(2)} = \text{_____}$$

$$I_3 = I_{3(1)} + I_{3(2)} = \text{_____}$$

Compare the above results to the measurements recorded in Table 10-2. According to superposition, the results should be the same.

_____

_____

11. Combine the results of Table 10-4 and Table 10-6.

$$V_{ab} = V_{ab(1)} + V_{ab(3)} = \underline{\hspace{3cm}}$$

$$V_{b} = V_{b(1)} + V_{b(3)} = \underline{\hspace{3cm}}$$

$$V_{bc} = V_{bc(1)} + V_{bc(3)} = \underline{\hspace{3cm}}$$

$$I_{1} = I_{1(1)} + I_{1(3)} = \underline{\hspace{3cm}}$$

$$I_{2} = I_{2(1)} + I_{2(3)} = \underline{\hspace{3cm}}$$

$$I_{3} = I_{3(1)} + I_{3(3)} = \underline{\hspace{3cm}}$$

Compare the above results to the measurements recorded in Table 10-3. According to superposition, the results should be the same.

_____

_____

PROBLEMS

12. Apply mesh analysis to calculate the loop currents in the circuit of Figure 10-1. Use the loop currents to determine currents I_1, I_2, and I_3.

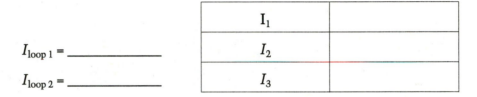

$I_{\text{loop 1}} =$ _____

$I_{\text{loop 2}} =$ _____

I_1	
I_2	
I_3	

13. Apply mesh analysis to calculate the loop currents in the circuit of Figure 10-2. Use the loop currents to determine currents I_1, I_2, and I_3.

$I_{\text{loop 1}} =$ _____

$I_{\text{loop 2}} =$ _____

I_1	
I_2	
I_3	

14. Use nodal analysis to calculate the node voltage V_b in the circuit of Figure 10-1. Your result should be very close to the measured value recorded in Table 10-2.

$V_b =$ _____

15. Use nodal analysis to calculate the node voltage V_b in the circuit of Figure 10-2. Your result should be very close to the measured value recorded in Table 10-3.

$V_b =$ _____

Thévenin's and Norton's Theorems (dc)

OBJECTIVES

After completing this lab, you will be able to
- determine the Thévenin and Norton equivalent of a complex circuit,
- analyze a circuit using the Thévenin or Norton equivalent circuit,
- determine the value of load resistance needed to ensure maximum power transfer to the load,
- measure the Thévenin (or Norton) resistance of a circuit using an ohmmeter,
- measure the Thévenin voltage of a circuit using a voltmeter,
- measure the Norton current of a circuit using an ammeter.
- describe how maximum power is transferred to a load when the load resistance is equal to the Thévenin resistance ($R_L = R_{Th}$).

EQUIPMENT REQUIRED

☐ Digital multimeter (DMM)
☐ dc power supply
Note: Record this equipment in Table 11-1.

COMPONENTS

☐ Resistors: 3.3-kΩ, 4.7-kΩ, 5.6-kΩ (1/4-W, 5% tolerance)
10-kΩ variable resistor

EQUIPMENT USED

Instrument	Manufacturer/Model No.	Serial No.
DMM		
dc Supply		

Table 11-1

TEXT REFERENCE

Section 9.2   THÉVENIN'S THEOREM
Section 9.3   NORTON'S THEOREM
Section 9.4   MAXIMUM POWER TRANSFER THEOREM

DISCUSSION

Thévenin's theorem:

Any linear bilateral network may be reduced to a simplified two-terminal network consisting of a single voltage source, E_{Th}, in series with a single resistor, R_{Th}. Once the original network is simplified, any load connected to the output terminals will behave exactly as if the load were connected in series with E_{Th} and R_{Th}.

Norton's theorem:

Any linear bilateral network may be reduced to a simplified two-terminal network consisting of a single current source, I_N, in parallel with a single resistor, R_N. A Thévenin equivalent circuit is easily converted into a Norton equivalent by performing a source conversion as follows:

$$R_N = R_{Th} \tag{11-1}$$

$$I_N = \frac{E_{Th}}{R_{Th}} \tag{11-2}$$

When a load is connected across the output terminals, the circuit will behave exactly as if the load were connected in parallel with I_N and R_N.

Maximum power transfer theorem:

Maximum power will be delivered to the load resistance when the load resistance is equal to the Thévenin (or Norton) resistance.

CALCULATIONS

1. Determine the Thévenin equivalent of the circuit of Figure 11-1. Sketch the equivalent circuit in the space provided below.

2. Determine the Norton equivalent of the circuit of Figure 11-1. Sketch the equivalent circuit in the space provided below.

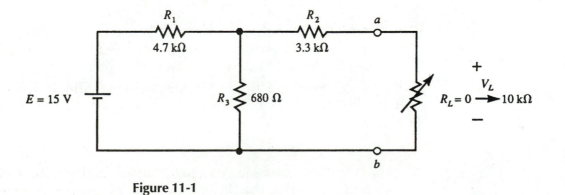

Figure 11-1

3. Calculate the minimum and the maximum voltage V_L which will appear across the load as R_L is varied between 0 and 10 kΩ. Enter the results below.

$V_{L\,(min)}$	
$V_{L\,(max)}$	

4. Calculate the minimum and the maximum load current I_L which will occur through the load as R_L is varied between 0 and 10 kΩ. Enter the results below.

$I_{L\,(min)}$	
$I_{L\,(max)}$	

5. Determine the value of load resistance for which maximum power will be transferred to the load.

R_L	

MEASUREMENTS

6. Assemble the circuit of Figure 11-1, temporarily omitting the load resistor R_L. Insert the DMM voltmeter across terminals a and b and measure the open-circuit voltage. This is the Thévenin voltage E_{Th}. Record the measurement here.

E_{TH}	

7. Insert the DMM ammeter between terminals a and b. Ensure that the ammeter is adjusted to measure the expected Norton current. Because the ammeter is effectively a short circuit, you are measuring the short-circuit current between terminals a and b. Record the measurement here.

I_N	

8. Remove the voltage source from the circuit and replace it with a short circuit. Place the DMM ohmmeter across terminals a and

b and measure the resistance between these terminals. This is the Thévenin resistance R_{Th}. Record the result here.

R_{Th}	

9. Reconnect the voltage source into the circuit. Connect the variable 10-kΩ resistor as a rheostat and insert it as the load resistance R_L. Adjust the variable resistor between its minimum and maximum values. Measure and record the maximum and minimum output voltage V_L.

$V_{L\,(min)}$	
$V_{L\,(max)}$	

10. Place the DMM ammeter in series with the load resistor R_L. Adjust the variable resistor between its minimum and maximum values. Measure and record the minimum and maximum load current I_L.

$I_{L\,(min)}$	
$I_{L\,(max)}$	

11. Connect the DMM voltmeter across the load resistor. Adjust R_L until the output voltage is exactly half of the Thévenin voltage measured in Step 6. When $V_L = E_{Th}/2$, the load resistor is receiving the maximum amount of power from the circuit. Carefully remove R_L from the circuit, ensuring that the rheostat is not accidently readjusted. Use the DMM ohmmeter to measure the value of R_L. Record the measurement here.

R_L	

CONCLUSIONS

12. Compare the measured value of Thévenin voltage E_{Th} in Step 6 to the calculated value of Step 1. Determine the percent variation.

   $E_{Th(theoretical)}$ = _____ percent variation = _____

13. Compare the measured value of Thévenin (or Norton) resistance R_{Th} in Step 8 to the calculated value of Step 1. Determine the percent variation.

   $R_{Th(theoretical)}$ = _____ percent variation = _____

14. Compare the measured value of Norton current I_N in Step 7 to the calculated value of Step 2. Determine the percent variation.

$I_{N(\text{theoretical})}$ = _____ percent variation = _____

15. An alternate method of determining the Thévenin (Norton) resistance is by applying Ohm's law to the Thévenin voltage and Norton current. Calculate the value of Thévenin resistance using the measured values of Thévenin voltage and Norton current.

$$R_{Th} = R_N = \frac{E_{Th}}{I_N} \qquad (11\text{-}3)$$

R_{Th}	

Compare this value to the actual measured value of Thévenin resistance of Step 8.

_____

_____

16. Compare the measured minimum and maximum voltage V_L to the calculated voltages determined in Step 3. Explain why there is a slight variation.

_____

_____

17. Compare the measured minimum and maximum load current I_L to the calculated currents determined in Step 4.

_____

18. When delivering maximum power to the load, how did the actual value of load resistance R_L compare the theoretical value calculated in Step 5?

_____

_____

Capacitors

OBJECTIVES

After completing this lab, you will be able to
- measure capacitance,
- verify capacitor relationships for series and parallel connections,
- verify that a capacitor behaves as an open circuit for steady state dc.

EQUIPMENT REQUIRED

☐ DMM
☐ Capacitance meter
☐ Variable dc power supply

COMPONENTS

☐ Capacitors: One each of 1 µF, 0.47 µF and 0.33 µF, non-electrolytic, 35 WVDC or greater
☐ Resistors: One each of 2.7-kΩ, 3.9-kΩ, and 10-kΩ, 1/4 W

EQUIPMENT USED

Instrument	Manufacturer/Model No.	Serial No.
DMM		
Capacitance meter[†]		
Power supply		

[†]Such as a DMM with capacitance measuring capability, an LCR meter (e.g., the BK Precision 878), a digital capacitance bridge, or an impedance bridge.

Table 12-1

TEXT REFERENCE

Section 10.1   CAPACITANCE
Section 10.7   CAPACITORS IN PARALLEL AND SERIES

DISCUSSION

A *capacitor* is a charge storage device and its electrical property is called *capacitance*. The more charge that a capacitor can store for a given voltage, the larger its capacitance. Capacitance, charge and voltage are related by the equation

$$C = \frac{Q}{V} \tag{12-1}$$

where C is the capacitance of the capacitor in *farads*, Q is the charge stored (in coulombs) and V is the voltage across its terminals. (Note: The farad is a very large quantity, and practical capacitors range in size from pF to hundreds of µF.) Because of its ability to store charge, a capacitor holds its voltage. That is, if you charge a capacitor, then disconnect the source, a voltage will remain on the capacitor for a considerable length of time. Dangerous voltages can be present on charged capacitors. For this reason, you should discharge capacitors before working with them.

For capacitors in parallel, the total capacitance is the sum of the individual capacitances. That is,

$$C_T = C_1 + C_2 + ... + C_N \tag{12-2}$$

For capacitors in series, the total capacitance may be found from

$$\frac{1}{C_T} = \frac{1}{C_1} + \frac{1}{C_2} + ... + \frac{1}{C_N} \tag{12-3}$$

Steady State Capacitor Currents and Voltages

Since capacitors consist of conducting plates separated by an insulator (called a dielectric), there is no conductive path from terminal to terminal. Thus, when a capacitor is placed across a dc source, except for a brief charging current, its current is zero. This means that a capacitor in steady state looks like an open circuit to dc.

When connected in parallel, the voltage across capacitances is the same. However, when connected in series, voltage divides in inverse proportion to the size of the capacitances: that is, the smaller the capacitance, the larger the voltage. For capacitors in series with a dc voltage source E connected across the string, Figure 12-3(a) (p. 98), the steady dc voltages on individual capacitors are related by

$$V_x = \left(\frac{C_T}{C_x}\right)E, \quad V_1 = \left(\frac{C_2}{C_1}\right)V_2, \quad V_1 = \left(\frac{C_3}{C_1}\right)V_3 \qquad (12\text{-}4)$$

and so on. (This is the voltage divider rule for capacitance.)

Measuring Capacitance

Measuring capacitance with a modern capacitor tester such as a DMM with a capacitance feature, or an LRC tester, or an automatic digital bridge is straightforward. The general procedure is

1. Short the capacitor's leads to discharge the capacitor. Remove the short.
2. Set the function selector to the appropriate capacitance range (if not autoranging), then connect the capacitor. Observe polarity markings on polarized capacitors if applicable. (This is not necessary on some testers.)
3. Read the capacitance value directly from the numeric readout.

With such testers, it takes only a few seconds to measure capacitance.

MEASUREMENTS

1. Carefully measure each capacitor and resistor and record their values in Tables 12-2 and 12-3.

	Nominal	Measured
C_1	1 µF	
C_2	0.47 µF	
C_3	0.33 µF	

	Nominal	Measured
R_1	2.7 kΩ	
R_2	3.9 kΩ	
R_3	10 kΩ	

Table 12-2   **Table 12-3**

2. a. Assemble each circuit of Figure 12-1 and measure capacitance C_T. Record results in Table 12-4.
   b. Verify the values measured in (a) by analyzing each circuit using the measured capacitor values from Table 12-2. Record calculated values in Table 12-4. How do they compare to the measured results?

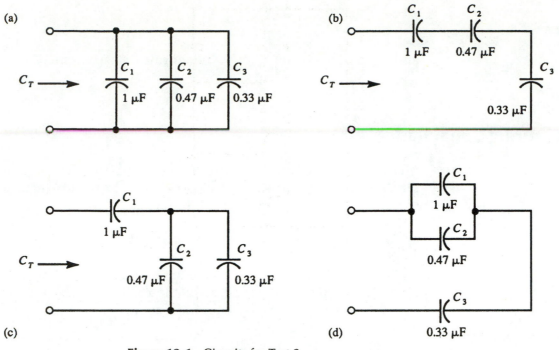

Figure 12-1   Circuits for Test 2

	Total Capitance C_T (µF)		
Circuit	Measured	Calculated	% of Difference
Figure 12-1(a)			
Figure 12-1(b)			
Figure 12-1(c)			
Figure 12-1(d)			

Table 12-4

3. a. Assemble the circuit of Figure 12-2. Set E = 18.0 V, measure V_1, V_2, and V_3 and record in Table 12-5.

	Measured	Computed
V_1		
V_2		
V_3		

Table 12-5

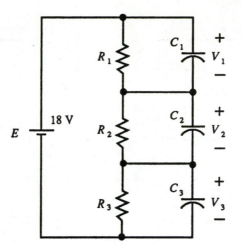

Figure 12-2   Circuit for Test 3

b. Verify the results of Table 12-5 by analyzing the circuit of Figure 12-2 and calculating voltages. Why do Equations 12-4 not apply here? Why do the capacitors not load the circuit?

PROBLEMS

4. For the circuit of Figure 12-3(a), compute the voltage across each capacitor and record in Table 12-6. Repeat for Figure 12-3(b).

	Figure 12-3(a)	Figure 12-3(b)
V_1		
V_2		
V_3		

Table 12-6

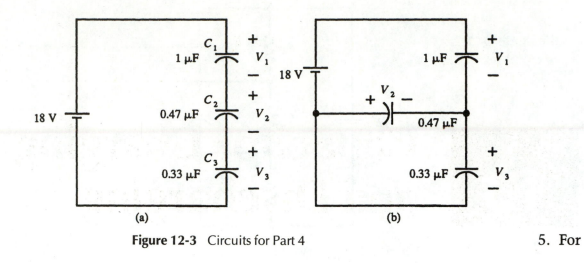

(a)   (b)

Figure 12-3   Circuits for Part 4

5. For the circuit of Figure 12-4, compute the voltage across the capacitor.

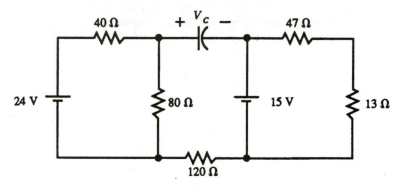

Figure 12-4   Circuit for Part 5

LAB

13

Capacitor Charging and Discharging

OBJECTIVES

After completing this lab, you will be able to
- measure capacitor charge and discharge times,
- confirm the voltage/current direction convention for capacitors,
- confirm the Thévenin method of analysis for capacitive charging and discharging,

EQUIPMENT REQUIRED

☐ DMM, VOM
☐ Variable dc power supply
☐ Function generator (optional)
☐ Oscilloscope (optional)

COMPONENTS

☐ Capacitors: 470-µF (electrolytic), 0.01-µF (non-electrolytic)
☐ Resistors: 10-kΩ, 20-kΩ, 39-kΩ, 47-kΩ, 1/4-W
☐ Switch: Single pole, double throw
☐ Stopwatch

Note to the Instructor

Part D of this lab may be run as an instructor demo if your students have not yet had instruction on the oscilloscope. If you want your students to perform an OrCAD PSpice or an Electronics Workbench analysis of this lab, use Lab #35, Part A.

EQUIPMENT USED

Instrument	Manufacturer/Model No.	Serial No.
DMM		
Power supply		
Function generator (optional)		
Oscilloscope (optional)		

Table 13-1

TEXT REFERENCE

Section 10.8   CAPACITOR CURRENT AND VOLTAGE
Section 11.1   INTRODUCTION
Section 11.2   CAPACITOR CHARGING EQUATIONS
Section 11.4   CAPACITOR DISCHARGING EQUATIONS
Section 11.5   MORE COMPLEX CIRCUITS

DISCUSSION

Capacitor charging and discharging may be studied using the circuit of Figure 13-1. When the switch is in position 1, the capacitor charges at a rate determined by its capacitance and the resistance through which it charges; when the switch is in position 2, it discharges at a rate determined by its capacitance and the resistance through which it discharges. This phenomenon of charging and discharging is important as it affects the operation of many circuits.

> **Caution**
> In this lab, you use electrolytic capacitors. Electrolytics are polarized and must be used with their + lead connected to the positive side of the circuit and their – lead to the negative side. An incorrectly connected electrolytic may explode.

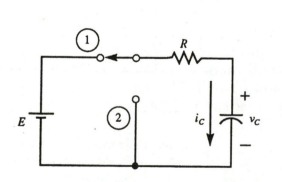

Figure 13-1   Circuit for studying capacitor charging and discharging

Voltage and Current During Charging

Consider Figure 13-2. At the instant the switch is moved to the charge position, current jumps from zero to E/R (since the capacitor looks like a short circuit at this instant). As the capacitor voltage approaches full source voltage, current approaches zero (since the capacitor looks like an open circuit to dc). During charging, voltage and current are given by

$$i_C = \frac{E}{R} e^{-t/RC} \tag{13-1}$$

$$v_C = E(1 - e^{-t/RC}) \tag{13-2}$$

The product RC is referred to as the *time constant* and is given the symbol τ. Thus,

$$\tau = RC \tag{13-3}$$

In one time constant, the capacitor voltage climbs to 63.2% of its final value while the current drops to 36.8% of its initial value. For all practical purposes, charging is complete in five time constants.

Voltage and Current During Discharging

The discharge curves are shown in Figure 13-3. When the switch is moved to position 2, the capacitor looks momentarily like a voltage

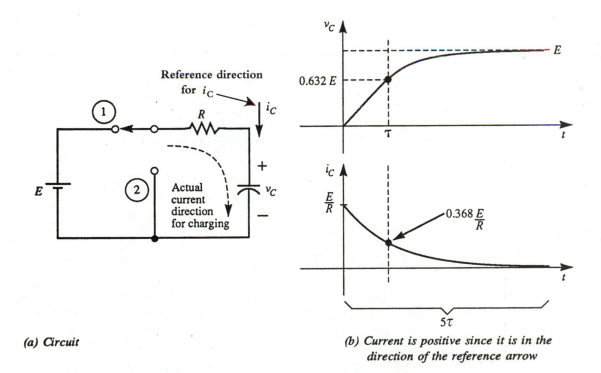

(a) Circuit

(b) Current is positive since it is in the direction of the reference arrow

Figure 13-2 Capacitor voltage and current during charging

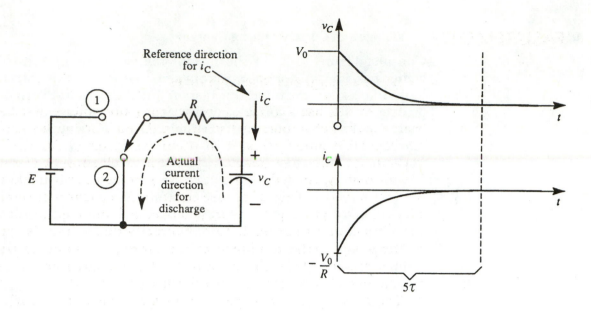

(a) Circuit

(b) Current is negative since it is opposite in direction to the reference arrow.

Figure 13-3   Capacitor voltage and current during discharging

source with value V_0 where V_0 is the voltage on the capacitor at the instant the switch is moved. (If the capacitor is fully charged, $V_0 = E$.) Current jumps from zero to $-V_0/R$, then decays to zero. (It is negative since it is opposite in direction to the reference as indicated in Figure 13-3a.) Voltage decays from V_0 to zero. Equations are

$$i_C = -\frac{V_0}{R}\,e^{-t/RC} \quad \text{or} \quad i_C = -\frac{E}{R}\,e^{-t/RC}$$

(13-4)

$$v_C = V_0 e^{-t/RC} \quad \text{or} \quad v_C = E e^{-t/RC}$$

(13-5)

Discharging takes five time constants. Although for this particular circuit charge and discharge resistances are the same, in general they are different. For the latter case, charge and discharge time constants will be different.

A Final Note

To properly observe capacitor charging and discharging, you need an oscilloscope. Since you may not have studied the oscilloscope at this time, we will examine the basic ideas in other ways. However, at the end of the lab, we have included an instructor demonstration (Part D) using an oscilloscope. Here, you will be able to observe capacitor charging and discharging directly on the screen.

MEASUREMENTS

PART A: Charge and Discharge Times

For Part A, use a VOM to observe capacitor voltage and a stopwatch to time charging and discharging. (Use a VOM with a sensitivity of 20,000 ohms per volt to minimize loading.) For the large capacitance value required here, you need an electrolytic capacitor. Unfortunately, you will find that your results agree only moderately well with theory. This is because electrolytic capacitors are far from ideal. (Electrolytic capacitors are polarized and thus require a dc voltage to maintain their oxide film. Normally this voltage is applied continuously. In this test however, there is no continuous polarizing voltage present. Under this condition, effective capacitance is larger than nominal capacitance. You will thus get values that are perhaps 15% to 20% higher than expected.) However, the results clearly verify the theory.

1. Assemble the circuit of Figure 13-1 with $R = 10$ kΩ, $C = 470$ μF and source voltage about 25 V. Place the switch in the charge position and wait for the capacitor to charge. Carefully adjust E until $V = 25$ V across the capacitor.

   a. Compute the time constant for the circuit.

   $\tau_{computed}$ = _____.

   b. Return the switch to the discharge position and wait for the capacitor to fully discharge. Move the switch to the charge position and using the stopwatch, time how long it takes for the capacitor voltage to reach 15.8 V (i.e., 63.2% of its final voltage). This is the measured time constant.

   $\tau_{measured}$ = _____ (charging).

   c. Hold the switch in the charge position until the capacitor voltage stabilizes at 25 volts. Now move the switch to the discharge position and time how long it takes for the voltage to drop to 9.2 V (i.e., to 36.8% of its initial value).

   $\tau_{measured}$ = _____ (discharging).

2. Change R to 20 kΩ and repeat Test 1.

   a. $\tau_{computed}$ = _____

   b. $\tau_{measured}$ = _____ (charging)

   c. $\tau_{measured}$ = _____ (discharging)

3. Describe how the results of Tests 1 and 2 confirm the theory.

_____

_____

PART B: Current Direction

To conform to the standard voltage/current convention, the + sign for voltage v_C must be at the tail of the current direction arrow i_C as indicated in Figure 13-1. This is obviously correct for charging as indicated in Figure 13-2. During discharge, the polarity of the voltage does not change. Thus, for discharging, the current direction arrow must also remain in the clockwise direction, even though we know that actual current is in the opposite direction as indicated in Figure 13-3. The interpretation is that for charging, current is in the same direction as the reference and hence, is positive, while for discharging, it is opposite to the reference and hence, is negative.

4. Add a DMM with autopolarity as in Figure 13-4. Connect the current input jack A to the positive side of the circuit as indicated. (Since the meter measures current _into_ terminal A, this connection will yield a positive value when current is in the reference direction and a negative value when opposite to the reference.) We now verify charge and discharge directions experimentally.

   a. Move the switch to charge and note the sign of the multimeter reading. (Don't try to read its value since the current is not constant, but is decaying to zero.) Sign _____. Thus, the actual

      direction of current is _____

   b. Move the switch to discharge and again note the sign. Sign

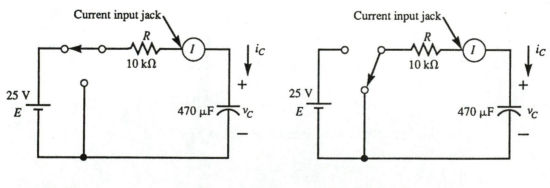

a) Charge   b) Discharge

Figure 13-4   Verifying the current direction convention

_____. Thus, the actual direction of current is

_____.

PART C: More Complex Circuits

For analysis, complex circuits can be reduced to their Thévenin equivalent. In this test, you will verify the Thévenin equivalent method.

5. a. Using Thévenin's theorem, determine the charge and discharge equivalent circuits for the circuit of Figure 13-5(a). Sketch in (b) and (c) respectively.

    b. From the Thévenin equivalents, compute the charge and discharge time constants.

    $\tau_{charge} = $ _____ . $\tau_{discharge} = $ _____ .

    c. For charging, compute capacitor voltage at $t = \tau_{charge}$.

    $v_C = $ _____ .

    d. Assume the switch has been in the charge position long enough for the capacitor to fully charge. Now move the switch to discharge and compute the voltage after one discharge time constant. $v_C = $ _____ .

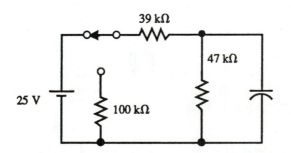

(a) Circuit. C = 470 μF   (b) Charge equivalent   (c) Discharge equivalent

Figure 13-5 Thévenin analysis

e. Assemble the circuit of Figure 13-5(a). Move the switch to charge and with the stopwatch, measure how long it takes for the voltage to reach the value determined in Test 5(c).

$\tau_{charge(measured)} =$ _____.

f. Allow the capacitor to fully charge, then move the switch to discharge and with the stopwatch, measure the time that it takes for the voltage to reach the value determined in Test 5(d).

$\tau_{discharge(measured)} =$ _____.

Discuss results. In particular, how well do the Thévenin equivalents represent the charging/discharging behavior of the capacitor in the real circuit?

PART D: Charging/Discharging Waveforms

Preliminary Note   Part D is set up as an instructor demonstration. (If you have already had instruction on the oscilloscope in this or another course, you may perform this part yourself.) The circuit is shown in Figure 13-6 (next page). A function generator simulates switching by applying a signal that cycles between source voltage and ground. A high rate of switching and a smaller time constant are used to give a waveform that can be viewed on the oscilloscope. Since a non-electrolytic capacitor is used, agreement between theory and practice will be much better than in previous tests.

6. a. Measure R and C, then assemble the circuit of Figure 13-6. Using the function generator, apply an input with 0.5 ms high and low times (i.e., $f = 1$ kHz) and adjust the input to 5 V as indicated in Figure 13-6. Set the oscilloscope to display only the charging voltage. Sketch below (next page) as Figure 13-7.

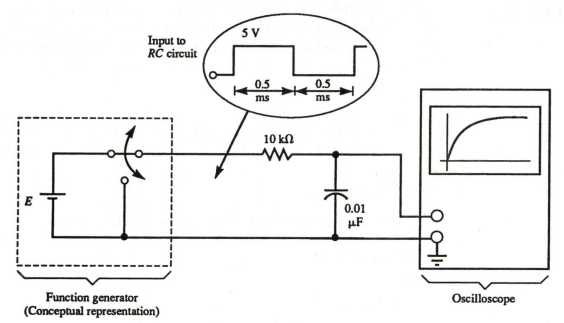

Input to
RC circuit

5 V

0.5
ms

0.5
ms

10 kΩ

0.01
μF

E

Function generator
(Conceptual representation)

Oscilloscope

Figure 13-6 Test set-up for Part D

b. Calculate the time constant using the measured *R* and *C*.

τ = _____.

c. From the scope screen, measure capacitor voltages at $t = \tau$, 2τ, 3τ, 4τ, and 5τ and tabulate.

d. Using circuit analysis techniques, compute voltages at these points. How well do they agree?

e. Display the discharge potion of the waveform and measure capitor voltages at $t = \tau$, 2τ, 3τ, 4τ, and 5τ and tabulate.

f. Repeat Part d) for the discharge values.

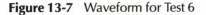

Figure 13-7 Waveform for Test 6

PROBLEMS

7. Assume the circuit of Figure 13-1 has a time constant of 100 μs. If R is doubled and C is tripled, calculate the new time constant.

8. For Figure 13-1, replace the wire between switch position 2 and common with a resistor R_2. If $R_2 = 4R$ and the capacitor takes 25 ms to reach full charge, how long will it take to discharge?

9. For Figure 13-8, the switch is closed at $t = 0$ s and opened 5 s later. The capacitor is initially uncharged.
    a. Determine the capacitor current i_C at $t = 2$ s.
    b. Determine the capacitor current i_C at $t = 7$ s.

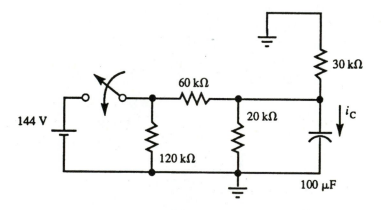

Figure 13-8   Circuit for Problem 9

NAME _____

DATE _____

CLASS _____

LAB 14

Inductors in dc Circuits

OBJECTIVES

After completing this lab, you will be able to
- determine steady state dc voltages and currents in an *RL* circuit,
- measure the inductive "kick" voltage in an *RL* circuit,
- measure the time constant of an *RL* circuit.

EQUIPMENT REQUIRED

☐ DMM
☐ Variable dc power supply
☐ Function generator
☐ Oscilloscope

COMPONENTS

☐ Resistors: 10-Ω, 82-Ω (two), 1-kΩ (1/4-W),
                      47-Ω, 220-Ω (1/2-W)
                      120-Ω (two, each 2-W),
☐ Inductors: One, 2.4-mH, powdered iron core, Hammond
                      Part #1534 or equivalent; one approximately
                      1.5-H, iron core.

Note to the Instructor

Parts C and D of this lab may be run as an instructor demo if your students have not yet had instruction on the oscilloscope. If you want your students to perform an OrCAD PSpice or an Electronics Workbench analysis of this lab, use Lab #35, Part B.

EQUIPMENT USED

Instrument	Manufacturer/Model No.	Serial No.
DMM or VOM (preferably DMM)		
Power supply		
Function generator		
Oscilloscope		

Table 14-1

TEXT REFERENCE

Section 13.2   INDUCED VOLTAGE AND INDUCTION
Section 13.7   INDUCTANCE AND STEADY STATE DC
Section 14.1   INTRODUCTION
Section 14.2   CURRENT BUILD UP TRANSIENTS
Section 14.4   DE-ENERGIZING TRANSIENTS

DISCUSSION

Induced voltage is determined by Faraday's law. For a coil with no magnetic saturation, induced voltage is directly proportional to the rate of change of current and is given by

$$v_L = L\frac{di}{dt} \tag{14-1}$$

where L is the inductance of the coil and di/dt is the rate of change of the current in the coil. The induced voltage opposes the change in current.

For inductances in series, total inductance is the sum of individual inductances. Thus,

$$L_T = L_1 + L_2 + ... + L_N \tag{14-2}$$

For inductances in parallel, total inductance may be found from

$$\frac{1}{L_T} = \frac{1}{L_1} + \frac{1}{L_2} + ... + \frac{1}{L_N} \tag{14-3}$$

Since real inductors have coil resistance, Equation 14-3 can only be used in practice if coil resistance is negligible.

Steady State and Transient Response

Steady State dc: As Equation (14-1) shows, voltage results only when current changes. Thus if current is constant (as in steady state dc), the voltage across an inductance is zero. Consequently, *to steady state dc, an inductance looks like a short circuit.*

Current Build up Transients: Simple current build up transients in *RL* circuits may be studied using the circuit of Figure 14-1(a). When the switch is closed, current in the inductor is given by

$$i_L = \frac{E}{R_1}(1 - e^{-R_1 t/L}) \tag{14-4}$$

which has a final steady state value of E/R_1 amps as shown in (b). Voltage is given by

$$v_L = Ee^{-R_1 t/L} \tag{14-5}$$

As Figures 14-1(b) and (c) show, at the instant the switch is closed, current is zero and full source voltage appears across the inductance. This means that *an inductor, with initial current of zero amps, looks like an open circuit.*

Current Decay Transients: Consider Figure 14-2(a) (next page). Assume that the current in the inductor at the instant the switch is opened is I_o. Since current cannot change instantaneously, *a current carrying inductance looks momentarily like a current source of I_o at the instant of switch operation.* Current then decays to zero according to

$$i_L = I_0 e^{-Rt/L} \tag{14-6}$$

where $R = R_1 + R_2$ for the circuit of Figure 14-2. Other voltages and currents can be obtained from this relationship using basic circuit principles. For example, the voltage v_2 across resistor R_2 is $-R_2 i_L$. Thus, it has the same shape as i_L but is negative as indicated in (c). (Multiplying $-R_2$ times Equation 14-6 yields $v_2 = -I_0 R_2 e^{-Rt/L}$ as indicated in Figure 14-2c.) Note that V_0 (which is equal to $-I_0 R_2$) can be many times larger than the source voltage.

The time constant of an RL circuit is given by

$$\tau = L/R \tag{14-7}$$

where R is the resistance through which the current builds or decays; for Figure 14-2, $\tau = L/R_1$ during current build up and $\tau = L/(R_1 + R_2)$ during current decay. Steady state is reached in 5τ where the appropriate τ (charge or discharge) must be used.

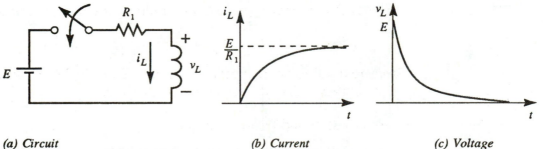

(a) Circuit   *(b) Current*   *(c) Voltage*

Figure 14-1 Current build up transients

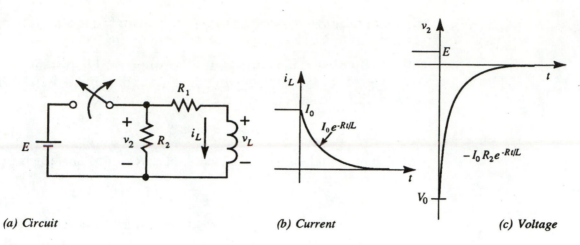

(a) Circuit   (b) Current   (c) Voltage

Figure 14-2 Current decay transients. $R = R_1+R_2$ for (a) and (b).

MEASUREMENTS

PART A: Steady State Voltages and Currents

1. Measure the resistance of each $120\,\Omega$ resistor and the resistance of the 2.4-mH inductor and record in Table 14-2.
2. a. Assemble the RL portion of the circuit of Figure 14-3. With the source disconnected, measure input resistance R_{in}.

   $R_{in(measured)} =$ _____

   b. Using circuit analysis techniques, compute R_{in}. $R_{in(computed)} =$

   _____. Compare to the measured value.

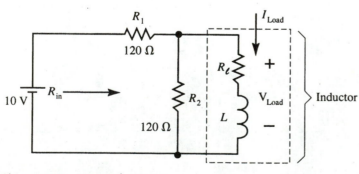

Figure 14-3 Circuit for Test 1

	Nominal	Measured
R_1	$120\,\Omega$	
R_2	$120\,\Omega$	
R_ℓ	xxxxx	

Table 14-2

c. Connect the source and measure voltage V_{Load} and I_{Load}.

$V_{Load(measured)}$ = _____ $I_{Load(measured)}$ = _____

d. Using circuit analysis techniques, solve for V_{Load} and I_{Load}. Compare to the measured values.

PART B: More Complex Steady State Circuits

For more complex circuits, use Thévenin's theorem to reduce portions of the circuit as necessary.

3. a. Measure each resistor for the circuit of Figure 14-4 and record in Table 14-3. Use the same inductor as in Figure 14-3.
    b. Determine the Thévenin equivalent of the circuit to the left of the inductor using circuit analysis techniques using the measured resistance values from Table 14-3.

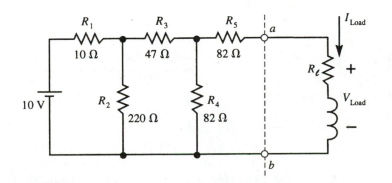

	Nominal	Measured
R_1	10 Ω	
R_2	220 Ω	
R_3	47 Ω	
R_4	82 Ω	
R_5	82 Ω	

Figure 14-4   Circuit for Test 3. Use resistor values measured in Table 14-3.

Table 14-3

c. Disconnect the inductor and measure the open circuit voltage across a-b. (This is E_{Th}.) $E_{Th(measured)}$ = _____.

d. Disconnect the source and replace it with a short circuit. Measure the resistance looking back into the circuit. (This is R_{Th}.)

$R_{Th(measured)}$ = _____

e. Compare the measured and computed values for E_{Th} and R_{Th}.

f. Remove the short, reconnect the source and inductor and measure I_{Load} and V_{Load}.

I_{Load} = _____ V_{Load} = _____

g. Using the Thévenin equivalent determined in (b), compute I_{Load} and V_{Load}. Compare to the measured values of (f).

PART C: Inductive "Kick" Voltage

In this part of the lab, you will look at the *inductive kick* that results when current in an inductor is interrupted. As noted in the previous lab, the only way to properly observe such a phenomenon is with an oscilloscope. If you have not yet covered the oscilloscope, this part of the lab may be run as an instructor demonstration.

Use the iron core inductor. (The inductor and resistor values used here are not critical. However, to see the results easily on the scope, you need a time constant of at least a few milliseconds. We first determine appropriate resistor values.)

4. a. Measure the dc resistance of the inductor. R_ℓ = _____.
   Consider Figure 14-5(a). Define $R_1' = R_1 + R_\ell$. Select a value for R_1 such that current E/R_1' is easily handled by your power supply. Now select an R_2 that is about five or six times larger than R_1' and such that the discharge time constant $L/(R_1+R_2+R_\ell)$ is a few milliseconds or more. (If you know L, you can compute τ; if you

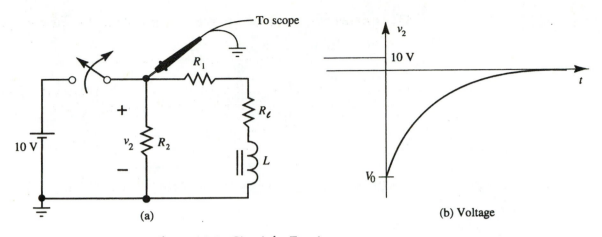

Figure 14-5   Circuit for Test 4

have no way of measuring L, experiment until you get a wave-form you can see.)

b. Assemble the circuit with these values and connect the scope probe across resistor R_2 to view the inductive kick voltage v_2. (The waveform will look like that shown in Figure 14-5(b). Voltage V_0 depends on the ratio of R_2 to R_1'. If $R_2 = 5\,R_1'$ and $E = 10$ V, then $V_0 = -50$ volts.) The waveform is a single shot event that is generated only when you open the switch. Set the scope triggering appropriately—e.g., *Norm* or *Single Sweep*, negative slope. With careful adjustment of triggering and the time base and with repeated operation of the switch, you should be able to observe the voltage spike of (b).

c. From the trace, measure the value of V_0. $V_0 =$ _____

d. Using the methods of Section 14.4 of the textbook, compute V_0 and compare to the measured value of (c). How do they compare?

PART D: Measuring the Time Constant

This part may also be run as a demo if necessary.

5. a. Measure the resistance of the 1-kΩ resistor and the resistance and inductance of the 2.4-mH inductor. (Call the 1-kΩ resistor R_S.) If you do not have any way to measure inductance, use the nominal value of 2.4 mH. Record below.

$R_1 =$ _____ $L =$ _____ $R_S =$ _____

   b. Assemble the circuit of Figure 14-6. The time constant for the circuit is $\tau = L/R_T$ where R_T is the sum of R_S, R and the output resistance R_{out} of your function generator. (This may be determined from the front panel of the generator. A typical output resistance of a function generator is 50 Ω.) Thus,

$\tau =$ _____.

   c. Set the function generator to the square wave mode at 40 kHz. (This provides more than ample time for current to build up and decay fully.) Adjust the signal generator and scope so that the amplitude of the v_S waveform is 5 grid lines. The waveform should look like that shown in Figure 14-6.

   d. Adjust the time base and triggering to get only the build up waveform on the screen. Sketch this waveform as Figure 14-7 (next page).

   e. Measure the time that it takes for the waveform to reach 63.2% of its final value. This is the measured time constant.

$\tau_{measured} =$ _____

Compare this value to the value computed in (b).

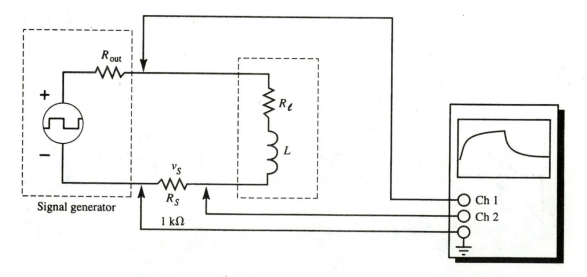

Figure 14-6 Current waveform for Test 5. The scope is triggered on Ch1 and the display is on Ch2. $R_T = R_{out} + R_\ell + R_S$.

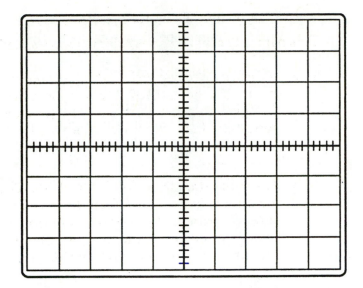

Figure 14-7   Waveform for Test 5

      f. Change triggering to get the decay waveform on the screen. Measure the time constant here and compare to that of (d).

$\tau_{measured} =$

PROBLEMS

    6. For the circuit of Figure 14-2, if E = 10 V, R_1 = 200 Ω, R_2 = 1200 Ω and L = 5 H,
      a. What voltage appears across the switch at the instant the switch is opened?
      b. How long will the transient last?

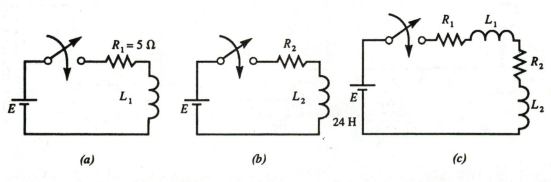

(a)   *(b)*   *(c)*

Figure 14-8

7. For the circuit of Figure 14-2, if $R_1 = 200\,\Omega$, $R_2 = 100\,\Omega$ and $L = 5$ H,
   a. What voltage appears across R_2 at the instant the switch is opened?
   b. How long will the transient last?
8. The circuit of Figure 14-8(a) takes 4 s to reach steady state, while that of (b) takes 12 s. The circuits of (a) and (b) are combined as in (c). How long will it take for the circuit of (c) to reach steady state?

The Oscilloscope (Part 1) Familiarization and Basic Measurements

OBJECTIVES

After completing this lab, you will be able to
- describe the operation and use of an oscilloscope,
- connect an oscilloscope to a circuit under test and select basic control settings,
- measure dc voltage,
- use an oscilloscope to observe time varying waveforms.

EQUIPMENT REQUIRED

☐ Oscilloscope
☐ Variable dc power supply
☐ Function generator
☐ DMM or VOM

Preliminary Note

Basic features of the oscilloscope are covered in *A Guide to Lab Equipment and Laboratory Measurements* at the beginning of this manual. You may wish to review this material before doing the lab.

EQUIPMENT USED

Instrument	Manufacturer/Model No.	Serial No.
DMM or VOM (preferably a DMM)		
Power supply		
Function generator		
Oscilloscope		

Table 15-1

DISCUSSION

The oscilloscope is the key test and measurement instrument used for studying time varying waveforms. Its main feature is that it displays waveforms on a screen; with an oscilloscope, you can view and study waveforms, measure ac and dc voltages and currents, frequency, period, phase displacement and so on. However, the oscilloscope is a fairly complex instrument and we therefore learn about it in stages. In this lab, we concentrate on operational procedures, front panel controls and a few basic measurements; in Labs 18 and 19, we look at more advanced measurement techniques. Later labs add more detail.

Connecting to the Circuit Under Test

The oscilloscope is connected to the circuit under test by means of a probe (or set of probes) as illustrated in Figure 15-1. The probe includes a measurement tip and a ground clip and connects to the oscilloscope via a flexible, shielded cable which is grounded at the oscilloscope. This ground serves as the reference point with respect to which all signals are measured. The shield helps guard against electrical noise pick up.

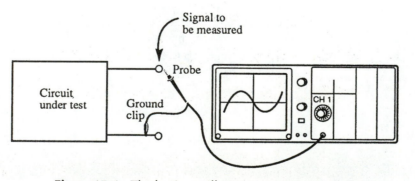

Figure 15-1   The basic oscilloscope measurement circuit

Probes may be "×1" or "×10". A "×10" probe contains a 10:1 voltage divider which attenuates the signal by a factor of 10; thus, when you use a "×10" probe, you have to multiply the scope readings by a factor of 10 to get the correct input (unless your scope automatically changes scales for you as some models do).

Front Panel Controls

Front panel controls permit you to control the operation of the oscilloscope. They may be grouped functionally as in Table 15-2.

Main Oscilloscope Controls According to Function			
Display	**Vertical**	**Horizontal**	**Triggering**
Intensity	Coupling (ac-gnd-dc)	Sec/Div	Coupling
Focus	Volts/Div	X-position	Source
Beam Finder	Y-position		Level
	Channel Select		Slope
			Mode

Table 15-2

A few of these are summarized below. Others will be introduced in later labs.

Coupling (ac-gnd-dc): Permits selection of coupling. When set to *dc*, the entire signal (ac plus any dc present) is displayed; when set to *ac*, dc signals are blocked by a capacitor and only ac is displayed; when set to *gnd*, the input is grounded. This permits establishing a *0-V base-line* (reference) on the screen.

VOLTS/DIV: This is the scope's vertical sensitivity control. It is a calibrated control that establishes how many volts each major vertical scale division represents. For example, when set for 10-V/DIV, each grid line represents 10 volts. Each channel has its own independent VOLTS/DIV control. A fine adjust control is also provided, but it is not calibrated. Each scale division is usually 1 cm.

Y-Position: This is the vertical position control. Each channel has its own control. It moves the trace up or down for easier observation. It is not calibrated.

Channel Select: Permits displaying Ch1, Ch2, both, or their sum or difference.

SEC/DIV: This is a calibrated control that selects how many seconds each major horizontal division represents. (It is calibrated in s, ms, and µs.) One control handles all channels. (A non-calibrated fine adjust is also provided.)

X-Position: Positions the trace horizontally. One control handles all channels.

Trigger Source: Selects the trigger source, e.g., Ch1, Ch2, an external trigger, or the ac line.

Trigger Level: Permits you to adjust the point on the trigger source waveform where you want triggering to start.

Trigger Slope: Selects whether the scope is to trigger on the positive or negative slope of the trigger source waveform.

Trigger Mode: Modes include *auto, normal* and *single sweep.* In the auto mode, the sweep always occurs, even with no trigger present; in the normal mode, a trigger must be present; in the single sweep mode, a trigger is required but only one sweep results. (Other modes may be provided but we will not consider them here.)

MEASUREMENTS

PART A: General Familiarization

If you have trouble getting a trace on the screen, check the intensity control; if it is set too low, the trace may be very faint or not visible. *(Caution: Never leave a bright spot on the screen.)* If adjusting the intensity does not locate the trace, proceed as follows: select Ch1, set triggering to auto, set SEC/DIV to mid range, press and hold the *beam finder* control, then adjust the vertical control to locate the trace.

Tests 1 to 3 are performed with no input applied to the oscilloscope.

1. Rotate the *focus* and *intensity* controls and note their effect. Adjust until you get a sharply focused trace at a comfortable viewing level.
2. Adjust the *vertical position* control and note its effect. Center the trace vertically on the center.
3. Adjust the *horizontal position* control and note its effect. Center the trace horizontally on the screen.
4. Connect a probe to Ch1 and set the channel selector to Ch1. Touch the probe tip to the *calibration test point* on the front panel. (Do not connect the ground clip.) Adjust the *VOLTS/DIV* control, the *SEC/DIV* control and the *trigger controls* until you get the calibration waveform on the screen. (It should be a square wave.)

PART B: Measuring dc Voltage with the Oscilloscope

5. a. Set the channel selector to Ch1 and use a ×1 probe. (Ensure the *VOLTS/DIV* switch for Ch1 is on *CAL.*) Set the trigger to *auto.* Move the *ac-gnd-dc* switch to *gnd* and center the trace. Return the coupling switch to *dc.* Voltage can be

Caution
The ground points on oscilloscopes, power supplies, and other equipment are generally tied to the electrical power system

(continues next page)

determined from the screen using the relationship $V = (deflection) \times (VOLTS/DIV\ setting)$.

b. Connect the probe as in Figure 15-2 and set *VOLTS/DIV* to 1 V. With the voltmeter, set the power supply to 2 V and note the deflection on the screen. From this deflection, compute the measured voltage. (It should equal the applied voltage.) Record in Table 15-3.

c. Now change *VOLTS/DIV* to 2 V, set $E = 4$ V and note the position of the trace. Enter data in Table 15-3 and compute *V*. Repeat for $E = 15$ V at 5 *VOLTS/DIV*.

d. Replace the probe with a ×10 probe. Using the oscilloscope, set the supply successively to 10 V, 15 V and 22.5 V. Record data, including the *VOLTS/DIV* settings that you choose. Compare to the meter reading.

ground through the U-ground pin on the electrical power outlet (plug-in). Since this connection ties all ground points together, you must be careful when connecting ground clips, as it is easy to inadvertently short out a component or even accidentally ground an output. While these grounds are required for safety reasons, they make poor signal paths. Therefore, be sure to use the ground clip supplied with the scope probe when making measurements.

Probe	Input Voltage	VOLTS/DIV Setting	Deflection	Voltage from Oscilloscope
×1	2 V	1-V		
×1	4 V	2-V		
×1	15 V	5-V		
×10				10 V
×10				15 V
×10				22.5 V

Table 15-3

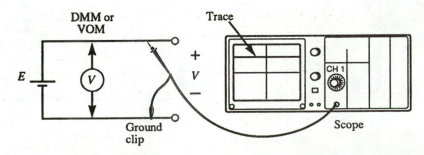

Figure 15-2 Circuit for Test 5

6. Move the input coupling switch to *ac*. Set the supply to the voltages of Table 15-3. What happens? Why?

_____

7. Return coupling to *dc*. Now set the trigger to *normal*. Note that the trace disappears. (This is because there is no trigger point in dc to start the sweep.) Return the trigger to *auto*.
8. Make sure your power supply output is floating. Reverse the probe connections to the power supply. Adjust the power supply voltage up and down. Note the deflection on the screen. Describe what happened.

_____

PART C Observing Waveforms

Replace the power supply with a function generator. Be sure to connect the ground of the oscilloscope to the ground of the generator. Set input coupling on the scope to *gnd* and center the trace. Change coupling to *ac*.

9. a. Set the function generator to a 2-kHz sine wave. On the oscilloscope, set the *VOLTS/DIV* switch to 1 V, the trigger to positive slope, and the time base to 0.1 ms/div. Adjust the output voltage of the generator until you get a nicely sized sine wave on the screen, then trim the frequency and adjust the horizontal position and trigger level controls to get one cycle of the waveform to fit between 5 horizontal grid lines. Sketch the waveform as Figure 15-3(a).

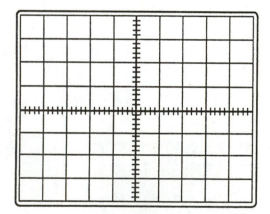

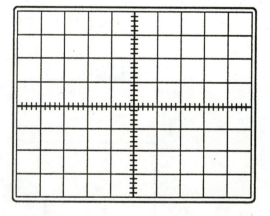

(a) Trigger slope set to positive   *(b) Trigger slope set to negative*

Figure 15-3   One cycle of a sine wave

b. Change the trigger slope to negative and repeat (a). Sketch the waveform as Figure 15-3(b).

c. Return the trigger slope to positive, then set input coupling to *dc*. (Make sure the dc offset on your function generator is set to zero.) What happens to the waveforms? (Compare to those of Figure 15-2.)

---

d. Return the coupling to *ac*. Set *f* = 500 Hz and change the time base to get 4 cycles on the screen (actually a bit more than four). Sketch here. Be sure to note the time base setting.

10. Repeat Steps 9(a) and (b) for a square wave and for a triangular wave. Sketch waveforms as Figures 15-4 and 15-5.

(a) Trigger slope set to positive   *(b) Trigger slope set to negative*

Figure 15-4   One cycle of a square wave

(a) Trigger slope set to positive   *(b) Trigger slope set to negative*

Figure 15-5   One cycle of a triangular wave

11. Set the generator to a 20 kHz sine wave. Adjust the controls to get one complete cycle on the screen. Record the time base setting.

    Time base setting _____

12. Take the time base control off *CAL* and vary it. Note and record what happens. Return the control to the *CAL* position.

    _____

PROBLEMS

13. A waveform is faint. What control should be adjusted?

    _____

14. A waveform is fuzzy. What control should be adjusted?

    _____

15. With a ×1 probe and *VOLTS/DIV* (vertical sensitivity) set at 5 V/div, a dc voltage moves the trace up 3.4 grid lines. What is the

    input voltage? _____

16. With a ×1 probe and the vertical sensitivity set at 10 V/div, a dc voltage moves the trace down by 1.2 grid lines. What is the input

    voltage? _____

17. With a ×10 probe and the vertical sensitivity control set at 5 V/div, a dc voltage moves the trace up 2.5 grid lines. What is the input voltage? (There are two possible answers, depending on your oscilloscope—see earlier note. Answer in terms of the scope you used in this lab.)

18. Using a ×1 probe and with the vertical sensitivity set to 2 V/div, a dc voltage moves the trace up by 2 and a half grid lines. The *VOLTS/DIV* fine adjust is used to bring the trace up to 3 grid lines high. What is the value of the input voltage? Why?

Basic ac Measurements: Period, Frequency, and Voltage (The Oscilloscope—Part 2)

OBJECTIVES

After completing this lab, you will be able to use an oscilloscope to

- measure period and frequency of an ac waveform,
- measure amplitude and peak-to-peak voltage,
- measure instantaneous voltage,
- determine the equation for a sinusoidal voltage from the oscilloscope readings.

EQUIPMENT REQUIRED

☐ Oscilloscope
☐ Signal or function generator

EQUIPMENT USED

Instrument	Manufacturer/Model No.	Serial No.
Oscilloscope		
Signal or function generator		

Table 16-1

TEXT REFERENCES

Measuring Period and Frequency: The period of a waveform is the length of one cycle. Since the horizontal scale of an oscilloscope is calibrated in seconds, you can measure the period T directly on the screen, then determine frequency from the relationship $f = 1/T$. For example, if the time base is set to 20 µs per division and one cycle is 4 divisions, then $T = 4(20 \text{ µs}) = 80 \text{ µs}$ and $f = 1/80 \text{ µs} = 12.5 \text{ kHz}$.

Measuring Voltage: An oscilloscope displays the instantaneous value of its input voltage. Thus, an oscilloscope may be used to measure peak voltage, peak-to-peak voltage, and indeed, the voltage at any point on a waveform. This voltage is measured in the same manner as dc voltage—you determine the deflection of the trace at that point, then multiply by the vertical sensitivity setting.

Equations for Sinusoidal Voltage from Oscilloscope Readings: Mathematically, the voltage at any point on a sine wave can be found from the equation

$$v = V_m \sin \alpha \qquad (16\text{-}1)$$

where α is the angular position on the cycle as indicated in Figure 16-1. If you know V_m, you can determine the voltage at any position by direct substitution into Equation 16-1. (Since one cycle represents 360°, one half cycle represents 180°, one quarter cycle represents 90°, and so on. The angular position at any other point can be determined by direct proportion.)

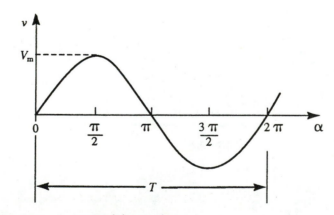

Figure 16-1   A sinusoidal waveform

Voltage as a Function of Time: Equation 16-1 may be rewritten as a function of time as

$$v(t) = V_m \sin \omega t \qquad (16\text{-}2)$$

where $\omega = 2\pi f = 2\pi/T$ and t is time measured in seconds. By measuring V_m and T on the screen, you can use Equation 16-2 to establish the analytic expression for any measured sinusoidal voltage.

Control Settings: Always select *VOLTS/DIV* and *SEC/DIV* settings to yield best results. For example, if you set the amplitude and time scales to spread a waveform over the screen, you can get more accurate measurements than if you compress the waveform into a small space. (Most settings here and in future labs will be left for you to select.) In addition, make sure that you have established an appropriate zero volt base line for each channel and that the *VOLTS/DIV* and *SEC/DIV* switches are set to their *CAL* (calibrate) positions.

A Final Note: The frequency scale of many signal generators is not very accurate and you may have to trim the frequency to get the desired period on the oscilloscope.

MEASUREMENTS

PART A: Measuring Period and Frequency

1. a. Connect the oscilloscope to the signal generator. Set the oscilloscope time base to 0.1 ms/div, coupling to *ac*, and the trigger slope to *positive*. Adjust the signal generator to obtain a sine wave that is 5 divisions in length. Thus, $T =$ _____ and $f =$ _____. Compare f to the frequency set on the generator dial.

    _____

    b. Repeat for a time base setting of 20 μs per division and two cycles in 8 divisions. $T =$ _____ and $f =$ _____. Compare to the frequency set on the signal generator dial.

    _____

PART B: Amplitude and Peak-to-Peak for a Sine Wave

2. For best results, set peak-to-peak amplitude rather than zero to

peak. Use the vertical control to position the waveform between grid lines. For this test, set the vertical sensitivity to 0.5 V/div and choose a frequency of 1 kHz.

a. Adjust the signal generator to yield a display of 8 grid lines peak-to-peak, centered vertically. What is peak-to-peak volt-

   age? $V_{\text{peak-to-peak}}$ = _____.

b. What is V_m? V_m = _____.

c. Sketch the waveform below with V_m and peak-to-peak voltages carefully labeled.

PART C: Instantaneous Value of a Sine Wave

3. a. Set the vertical sensitivity control to 1 V/div, the time base to 0.1 ms/div and obtain a waveform on the screen with a cycle length of exactly 8 divisions. Adjust the signal generator for a peak-to-peak display of 8 grid lines. Sketch the waveform as Figure 16-2 (next page). Label the vertical and horizontal axes in volts and ms.

   b. From the screen, measure the voltage at 0.1 ms intervals and record in Table 16-2.

   c. How many degrees does each 0.1 ms division represent?

   _____.

   Record the value of a for each value of t shown in Table 16-2.

   d. Using Equation 16-1, verify each entry in the table. (Show a few sample calculations below.)

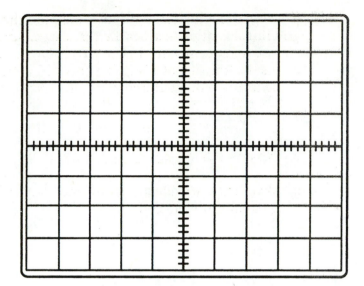

Figure 16-2   Measured waveform for Test 3

		Voltage	
t (ms)	a (deg)	Measured	Computed
0			
0.1			
0.2			
0.3			
0.4			
0.5			
0.6			
0.7			
0.8			

Table 16-2

PART D: The Equation for Sinusoidal Voltage

4. a. For the waveform of Figure 16-2, determine ω.

   $\omega =$ _____

   b. Using the measured values of ω and V_m, write the equation for the voltage in the form of Equation 16-2.

   $v(t) =$

c. Using this equation, compute v at $t = 50\,\mu s$ and $t = 150\,\mu s$. Show details here.

PROBLEMS

5. With the time base set to *0.5 µs/DIV*, four cycles of a waveform occupies 10 divisions. What is the period and the frequency of the waveform?

Period _____ Frequency _____

6. With the *VOLTS/DIV* set to 2 and a ×1 probe, a waveform has an amplitude of 2 and a half grid lines. The *VOLTS/DIV* fine adjust is used to bring the amplitude up to 3 grid lines high. What is the amplitude of the input voltage?

Amplitude _____

7. Given $v(t) = 100 \sin 377t$,

a. What is the value of V_m? $V_m =$ _____

b. What are the frequency and period?

$f =$ _____ $T =$ _____

c. Compute the voltage at $t = 20$ ms. Sketch the waveform and show where $t = 20$ ms occurs on the waveform.

8. Determine the equation $v(t)$ for the voltage depicted in Figure 16-3.

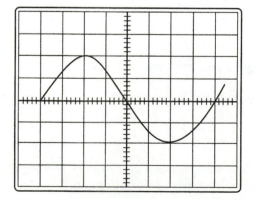

Vertical = 5V/div
Time base = 0.1 µs/div

Figure 16-3   Waveform for Question 8

AC Voltage and Current (The Oscilloscope— Part 3, Additional Measurement Techniques)

OBJECTIVES

After completing this lab, you will be able to use an oscilloscope to
- measure rms values for sinusoidal voltage,
- measure superimposed ac and dc voltages,
- measure ac current using a sensing resistor,
- display two waveforms simultaneously on a dual channel oscilloscope,
- measure phase displacement with a dual channel oscilloscope,
- measure voltage using differential measurement techniques.

EQUIPMENT REQUIRED

☐ Dual channel oscilloscope
☐ Signal or function generator
☐ DMM and VOM

COMPONENTS

☐ Resistors: 82-Ω, 100-Ω, 150-Ω, 1-kΩ, 2-kΩ, 3.3-kΩ, 5.6-kΩ, 6.8-kΩ, 10-kΩ (all 1/4-W)
☐ Capacitor: 0.01-μF
☐ Battery: 1.5-V

EQUIPMENT USED

Instrument	Manufacturer/Model No.	Serial No.
DMM		
VOM		
Dual Channel oscilloscope		
Signal or function generator		

Table 17-1

TEXT REFERENCE

Section 15.6 VOLTAGES AND CURRENTS AS FUNCTIONS OF TIME

Section 15.9 EFFECTIVE VALUES

DISCUSSION

Measuring the rms Voltage of a Sine Wave. An oscilloscope may be used to determine the rms value of a sinusoidal voltage since, for a sine wave, V_{rms} = 0.707 V_m and V_m can be measured directly on the screen. One of the reasons you might do this is convenience—if you have a signal already displayed on the screen, there is no need to connect a voltmeter. However, a more fundamental reason has to do with frequency—most meters have very limited frequency ranges. For example, VOMs can typically measure to only 100 kHz, while most common DMMs can measure to only a few kHz. (However, check your meter, as some DMMs are limited to less than a kHz, while others can measure to a few MHz.) On the other hand, even inexpensive oscilloscopes can measure to the tens of MHz, while top of the line units can measure to hundreds of MHz, or even to the GHz range with digital units.

Measuring Current. Oscilloscopes can also be used to measure current, although not directly. There are, however, two indirect ways in which you can measure current. The first is to insert a known resistor (sometimes called a *sensing resistor*) into the circuit, measure the voltage across it using an oscilloscope, then use Ohm's law to compute current. (This method is inexpensive and widely used. The other method is to use a *current probe*, but few introductory courses have access to equipment of this type so we won't consider it.)

The current sensing resistor approach is based on Ohm's law. For a purely resistive circuit, $v = Ri$. Thus, if voltage is sinusoidal, current will be sinusoidal also and vice versa—that is, v and i are in phase. Therefore, if voltage v is

$$v = V_m \sin \omega t$$

then $\qquad\qquad i = I_m \sin \omega t$

Current can thus be determined by measuring V_m and computing I_m from $I_m = V_m/R$. If the rms value of current is needed, it may be determined from the equation $I_{rms} = 0.707\ I_m$.

Dual-Channel Measurements. With a dual-channel oscilloscope, you can display two waveforms simultaneously. This permits you to determine phase relationships between signals, compare wave shapes, and so on.

Differential Voltage Measurements. Sometimes you need to measure the voltage across a component where the normal technique of placing the probe tip at one end and the ground clip at the other end shorts out part of the circuit. For problems such as this, *differential measurement* can be used. In Part E of this lab, you learn how to make such measurements.

MEASUREMENTS

PART A: RMS Values and the Frequency Response of ac Meters

We begin with a look at the frequency response of various instruments. Here, we compare the ability of the oscilloscope, DMM, and VOM to measure voltage at different frequencies.

1. a. Assemble the circuit of Figure 17-1. Adjust the signal generator to a 100 Hz sine wave with 12 V peak-to-peak (i.e., $V_m = 6$ V. Thus, $V_{rms} = 0.707 \times 6 = 4.24$ V. This is recorded in Table 17-2 (next page) as *Actual rms*.) Now measure and record the rms voltage using the DMM and the VOM.

    b. Repeat step (a) at the other frequencies indicated in Table 17-2.

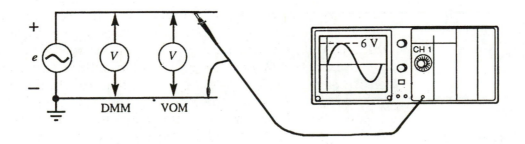

Figure 17-1   Circuit for Test 1

Frequency	Scope Reading	Actual rms	DMM reading	VOM reading
100 Hz	6 V	4.24 V		
1000 Hz	6 V	4.24 V		
10000 Hz	6 V	4.24 V		
100 kHz	6 V	4.24 V		
1 MHz	6 V	4.24 V		

Table 17-2

c. What conclusion do you draw from the data of Table 17-2?

_____

_____

PART B: Superimposed ac and dc

2. Add the 1.5-V battery to the circuit as in Figure 17-2(a). Set the signal generator to a 100-Hz sine wave. Select *ac* coupling (to temporarily block the dc component while you set the ac component) and adjust the output of the generator to V_m = 2 V (i.e., 4 V p-p). Return coupling to *dc*.
   a. You now have superimposed ac and dc voltages. Sketch as Figure 17-2(b).
   b. Compute the rms value for this waveform from $V = \sqrt{V_{dc}^2 + V_{ac}^2}$ where V_{dc} is the dc component of the waveform and V_{ac} is the rms value of its ac component.

   $V =$ _____.

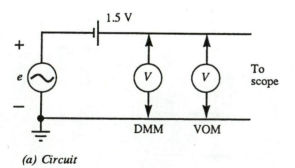

(a) Circuit

(b) Waveform

Figure 17-2   Circuit for Test 2

c. Measure the voltage using the meters.

DMM reading _____ VOM reading _____

d. Discuss the results of (b) and (c). In particular, why do the two meters not yield the rms value of the waveform? What type of meter is needed?

_____

_____

_____

PART C: Dual-Channel Measurements

We now learn how to display two waveforms simultaneously.

3. Assemble the circuit of Figure 17-3. Set the channel select switch to *alt* and trigger mode to *auto*. (*Alt* lets you display both Ch1 and Ch2 simultaneously.)
    a. Establish the 0-V baseline for Ch1 by moving its *ac-gnd-dc* switch to *gnd* and adjust its vertical position control until the trace is centered. Repeat for Ch2. (The traces should now be superimposed.) Return both Ch1 and Ch2 to the *ac* position and set triggering to Ch1, positive slope. Set *SEC/DIV* to 50 µs/div.
    b. Set the signal generator to a 2.5-kHz sine wave with $V_m = 3$ V (i.e., 6 V p-p). (You should now have two sine waves on the screen.) Select an appropriate setting for Ch2 and trim the vertical position controls if necessary to ensure that both waveforms are centered about the horizontal axis.
    c. Set the time base to 20 µs/div, and using the time base variable control if necessary, spread one half cycle of the reference waveform (Ch1) over the entire screen. (This results in each major scale division representing 18°.) Now measure the dis-

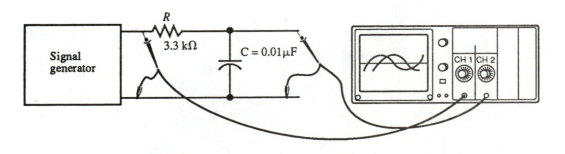

Figure 17-3 Circuit for Test 6. Use $R = 3.3$-kΩ and $C = 0.01$ µF.

placement between the waveforms at their zero-crossover points and convert this displacement to degrees.

d. As you will learn later, the theoretical displacement is $\theta = \tan^{-1}(1/\omega RC) - 90°$. Compute θ and compare to the result of (c).

e. Measure the magnitudes of each waveform. Taking the voltage of Ch1 as reference, sketch the waveforms in this space.

f. Write the equations for the two voltages, using the values measured above.

$v_1(t) =$

$v_2(t) =$

g. Set $f = 10$ kHz and again measure displacement. Using the equation of Test 3(d), compute displacement and compare it to the measured value.

h. Repeat Test (f) at a frequency of 18 kHz.

PART D: Current Measurement with an Oscilloscope

Consider Figure 17-4(a). The load current I_L is given by $I_L = E/R_L$ where E and I_L are the rms values of the source voltage and load current respectively. To measure this current using an oscilloscope, add a sensing resistor R_S as in Figure 17-4(b), measure the voltage across it, convert to rms, then compute current as $I_L' = V_S/R_S$. (As long as R_S is small compared to R_L, the accuracy will be good.) This method is valid regardless of waveform.

4. a. Accurately measure the 100-Ω sensing resistor and R_L. Set up the circuit of Figure 17-4(b). (Note the placement of the sensing resistor. With the resistor placed as shown, you can connect both ground clips of the oscilloscope directly to the ground of the signal generator without fear of ground problems.) Use a 100-Hz sinusoidal source. Using Ch1, set E_m to 3 V (i.e., 6 V p-p).

    b. Measure the voltage across the sensing resistor using Ch2, then convert to rms.

    $V_S =$ _____ (rms)

    c. Using Ohm's law, determine the rms value of the measured load current

    $I_L' = V_S/R_S =$ _____ (mA rms)

    Verify this current by comparing it to that measured by the meter.

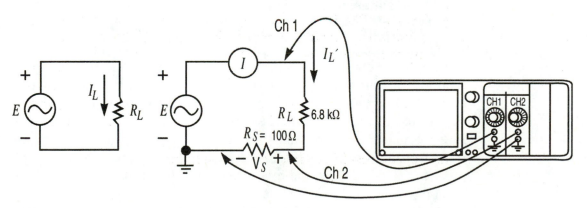

(a) Circuit   (b) Measuring current. Here, $I_L' \approx I_L$

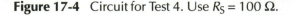

Figure 17-4   Circuit for Test 4. Use $R_S = 100\ \Omega$.

d. The load current that you are trying to measure is $I_L = E/R_L$, where E is the rms value of the source voltage. Using the measured value of R_L, compute I_L using this formula and compare to the current determined by the sensing resistor approach in (c).

5. Replace R_L of Figure 17-4 with the network of Figure 17-5 and repeat steps 4(b) to 4(d).

$V_S = $ _____ (rms voltage across R_S); $I_L' = V_S/R_S = $ _____

$I_L = $ _____ (the meter reading); $I_L = $ _____ (Determined by circuit analysis)

How do the results compare?

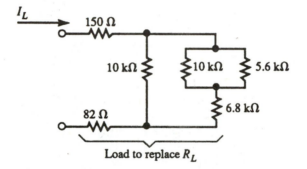

Figure 17-5   Circuit for Test 5. Replace R_L with this network.

PART E: Differential Measurements

Consider Figure 17-6 (next page). Suppose you want to measure the voltage across resistor R_1 using an oscilloscope. As a first thought, you might try the connection shown. With this connection, however, the ground lead shorts out R_2. Alternatively, you might try reversing the probe tip and ground clip connections. However, this shorts the source output to ground. Thus, neither approach is usable. (Of course, you might consider isolating the scope or the source from ground, but this is not desirable either, as you lose the earth safety ground.) A better approach is to use differential measurement as illustrated in Figure 17-7 (next page).

In Figure 17-7, Ch1 measures the voltage from point a to ground while Ch2 measures the voltage from point b to ground. The oscilloscope has a mode (its *differential mode*) that permits you to display Ch1 minus Ch2. This yields a display of v_{ab}, the voltage between points a and b.

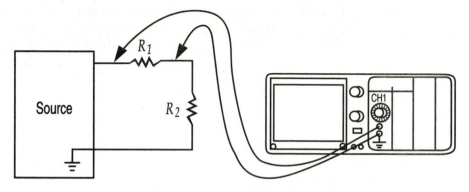

Figure 17-6 An incorrect way to view the voltage across R_1

6. a. Assemble the circuit of Figure 17-7. Set the signal generator to a 1 kHz sine wave.
   b. Set *VOLTS/DIV* to the same value for both channels and ensure that they are in the *CAL* position. Center the traces as in Step 3(a). Return the *ac-gnd-dc* switch to *ac* for both channels.
   c. Select Ch1 and set the input voltage to 3 V (i.e., 6 V p-p).
   d. Select the *difference mode* for your oscilloscope. (Details differ for different scopes. For example, some scopes require that you invert Ch2 and add it to Ch1. Check your scope's manual for details or ask your instructor.)
   e. You should now have a display of v_{ab} on your screen. Since $R_1 = 2/3\ R_T$, v_{ab} should be 2/3 of the source voltage. Verify that it is.
7. We will now look at what happens if you had tried to use the circuit of Figure 17-6 to measure the voltage across R_1.
   a. Consider again the circuit of Figure 17-7. Disconnect Probe 2 (i.e., Ch2) and set the channel select back to Ch1. You should see 3 V (i.e., 6 V p-p) on the screen since Ch1 is still measuring full source voltage.
   b. Move the ground clip of Probe 1 to point *b* and note the scope display.
   c. In Test 7(b), your probe is connected from *a* to *b*; thus, you are

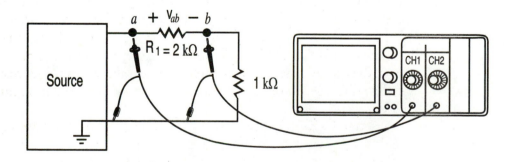

Figure 17-7 The correct way to view the voltage using differential measurement

measuring v_{ab}. However, it is different from v_{ab} you measured in Test 6(e). Briefly discuss what you are now seeing.

PROBLEMS

8. Given: $v_1(t) = 100 \sin \omega t$ and $v_2(t) = 80 \sin(\omega t + 30°)$ displayed on a scope screen with v_1 as reference. If $f = 100$ Hz, sketch the oscilloscope trace in the space below as Figure 17-8(a).

9. Repeat Question 8 (as Figure 17-8(b)) if v_2 is the reference waveform.

Figure 17-8   (a) Waveform for Question 8   (b) Waveform for Question 9

LAB

18

Inductive Reactance

OBJECTIVES

After completing this lab, you will be able to
- measure phase difference between voltage and current in an inductance,
- measure inductive reactance and verify theoretically,
- determine the effect of frequency on inductive reactance.

EQUIPMENT REQUIRED

☐ Dual channel oscilloscope
☐ Signal or function generator
☐ VOM (two)

COMPONENTS

☐ Resistors: 10-Ω, 100-Ω, 1/4-W
☐ Inductors: 2.4-mH, two required. (Hammond Part #1534 or equal. This inductor has a very small resistance which is necessary to approximate an ideal inductor.)

EQUIPMENT USED

Instrument	Manufacturer/Model No.	Serial No.
VOM 1		
VOM 2		
Dual channel oscilloscope		
Signal or function generator		

Table 18-1

TEXT REFERENCE

Section 16.5   INDUCTANCE AND SINUSOIDAL AC

DISCUSSION

Inductance and Sinusoidal ac. When an ideal inductance is connected to a sinusoidal voltage source, current lags voltage by 90° as illustrated in Figure 18-1. Thus, if

$$v_L = V_m \sin \omega t \qquad (18\text{-}1)$$

$$\text{then } i_L = I_m \sin(\omega t - 90°) \qquad (18\text{-}2)$$

$$\text{where } I_m = V_m/X_L \qquad (18\text{-}3)$$

The quantity X_L is termed *inductive reactance* and is given by the formula

$$X_L = \omega M \text{ ohms} \qquad (18\text{-}4)$$

where $\omega = 2\pi f$ rad/s. Inductive reactance represents the opposition that the inductance presents to current and is directly proportional to the product of inductance and frequency. Thus, the higher the frequency, the greater the opposition.

Effective (rms) Voltage and Current Relationships. In practice, we usually use effective values rather than peak values. Their ratios are the same, however. Thus reactance can also be expressed as

$$X_L = V_L/I_L \text{ ohms} \qquad (18\text{-}5)$$

where V_L and I_L are the rms values of v_L and i_L respectively. (Note that V_L and I_L are the values that you read on meters, while V_m and I_m are the values that you read on the oscilloscope.)

Practical Inductors. In reality, inductors have resistance as well as inductance—see Figure 18-2(a). However, if XL is large compared to

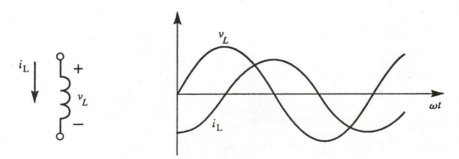

Figure 18-1   Voltage and current for an inductance

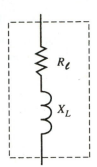

R_ℓ

X_L

(a) A real inductor

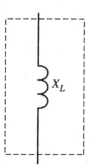

X_L

(b) Approximation when R_t has negligible effect.

Figure 18-2 The ideal inductor approximation

R_ℓ, we can neglect resistance and use the ideal model of (b). This means the simple formulas above apply. (In this lab, we will use an inductance where this approximation is very good at higher frequencies. However, at lower frequencies, the error in the approximation starts to show up. You will have a chance to see how good the approximation is at these lower frequencies.)

MEASUREMENTS

PART A: Phase Relationships for Inductance

Consider Figure 18-3. Voltage v_S across R_S is in phase with inductor current i_L. Provided circuit resistance is small compared to X_L, source voltage is approximately equal to the v_L. Thus, the phase difference between Ch1 and Ch2 is approximately equal to the angle between v_L and i_L. It should be close to 90°.

1. a. Measure coil resistance and inductance and the resistance of the sensing resistor, then assemble the circuit of Figure 18-3.

$L =$ _____ $R_\ell =$ _____ $R_S =$ _____

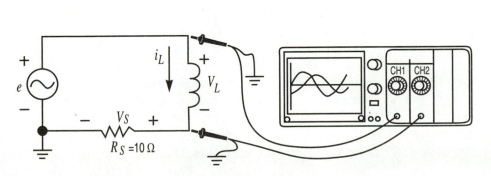

Figure 18-3 Circuit resistance is very small. Therefore, $v_L \approx e$.

b. Use a 40 kHz sine wave. Sketch the waveforms below. Label Ch1 as v_L and Ch2 as i_L. Using the procedure of Lab 17, determine the phase shift and indicate it on the diagram. How close is it to the theoretical value of 90°? Vary the frequency a few kHz and observe the shift. Is there any change? Based on this observation, state in your own words the phase relationship between voltage and current for an inductance.

c. The inductor will behave approximately as an ideal inductor provided $X_L \gg R$ where $R = R_\ell + R_s$. Check the approximation at 40 kHz. Use the measured R_ℓ and L determined in Test 1(a).

PART B: Magnitude Relationships for Inductance

Preliminary Notes. For the remaining portions of the lab, you need a pair of VOMs or DMMs with a frequency response up to 10 kHz. (Most VOMs can handle this range easily, but many low cost DMMs cannot measure much above 1 kHz. To be sure, check your meter manual.) If you do not have suitable meters, make measurements with the oscilloscope.

Since reactance depends on frequency, accurate frequency settings are essential. However, the frequency dial on many signal generators is not accurately calibrated. Thus, you may need the oscilloscope to set the frequency.

2. a. Replace resistor R_S of Figure 18-3 with an accurately measured 100-Ω resistor.

   $R_S = \underline{\hspace{2cm}}$

   b. Set the signal generator to a 10-kHz sine wave, 6 V peak to peak. With a meter (or the oscilloscope if necessary), carefully measure the voltage V_L across the inductor and voltage V_S across R_S. Calculate current I_L using Ohm's law and the measured value of V_S.

   $V_L = \underline{\hspace{2cm}}$ $V_S = \underline{\hspace{2cm}}$ $I_L = \underline{\hspace{2cm}}$

   c. Calculate the reactance of the inductor using the measured V_L and I_L.

   $X_{L(measured)} = V_L/I_L = \underline{\hspace{2cm}}$

   d. If possible, measure L using a bridge or RLC meter. (Otherwise, use the nominal value.) Compute X_L using Equation 18-4. Compare to the measured value of Test 2(c).

   $L = \underline{\hspace{2cm}}$ $X_{L(computed)} = \underline{\hspace{2cm}}$

3. a. Add the second inductor in series and determine total reactance using the same procedure that you used in Test 2(b) and (c).

   $V_L = \underline{\hspace{2cm}}$ $s = \underline{\hspace{2cm}}$

   $I_L = \underline{\hspace{2cm}}$ $X_{T(measured)} = \underline{\hspace{2cm}}$

   b. If possible, measure L for the second inductor and compute L_T for the series combination. $L_T = \underline{\hspace{2cm}}$. Now use Equation

18-4 to determine $X_{T(computed)}$. $X_{T(computed)} =$ _____. Compare to the measured value of 3(a).

4. a. Connect both inductors in parallel and determine their equivalent reactance using the procedure that you used in Test 2(b) and (c).

$V_L =$ _____ $V_S =$ _____

$I_L =$ _____ $X_{eq(measured)} =$ _____

    b. Compute L_{eq} for the parallel combination. $L_{eq} =$ _____.

Now use Equation 18-4 to determine $X_{eq(computed)}$. $X_{eq(computed)} =$

_____. Compare to the measured value of 4(a).

PART C: Variation of Reactance with Frequency

5. a. Using one of the 2.4-mH inductors, measure V_L and V_S and compute reactance (using the procedure of Test 2) at each of the frequencies listed in Table 18-2. Record as $X_{L(measured)}$.

    b. Using Equation 18-4, compute reactance at each frequency and record as $X_{L(computed)}$.

c. Using the graph sheet of Figure 18-4, plot measured and computed reactances versus frequency. Label each plot.
d. The plots should pass through $X_L = 0\ \Omega$ at $f = 0$ Hz. Do they? (Extrapolate both plots to find out.)
e. Analyze the results. That is, comment on how well the measured values agree with the computed values. Is the agreement poorer as the frequency gets lower? If so, why?

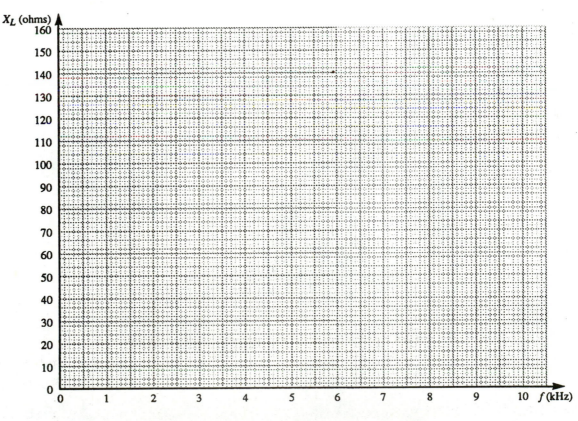

Figure 18-4   Plot of inductive reactance versus frequency

PROBLEMS

6. Considering ideal inductors, if you double inductance and triple frequency, what happens to current if the applied sinusoidal voltage remains the same?

f (kHz)	V_L	V_s	I_L	$X_{L(measured)}$	$X_{L(computed)}$

Table 18-2

NAME _____

DATE _____

CLASS _____

Capacitive Reactance

OBJECTIVES

After completing this lab, you will be able to
- measure phase difference between voltage and current in a capacitance,
- measure capacitive reactance and verify theoretically,
- determine the effect of frequency on capacitive reactance.

EQUIPMENT REQUIRED

☐ Dual channel oscilloscope
☐ Signal or function generator
☐ DMM (two)

COMPONENTS

☐ Resistors: 10-Ω, 220-Ω, 1/4-W
☐ Capacitors: 1.0-μF (two), non-electrolytic

EQUIPMENT USED

Instrument	Manufacturer/Model No.	Serial No.
DMM #1		
DMM #2		
Dual channel oscilloscope		
Signal or function generator		

Table 19-1

TEXT REFERENCE

Section 16.6   CAPACITANCE AND SINUSOIDAL AC

DISCUSSION

Capacitance and Sinusoidal ac: When an ideal capacitance is connected to a sinusoidal voltage source, current leads by 90° as illustrated in Figure 19-1. Thus, if

$$v_C = V_m \sin \omega t \qquad (19\text{-}1)$$

$$\text{then } i_C = I_m \sin(\omega t + 90°) \qquad (19\text{-}2)$$

$$\text{where } I_m = V_m/X_C \qquad (19\text{-}3)$$

The quantity X_C is termed *capacitive reactance* and is given by the formula

$$X_C = 1/\omega C \text{ ohms} \qquad (19\text{-}4)$$

where $\omega = 2\pi f$ rad/s. Capacitive reactance represents the opposition that a capacitor presents to current and is inversely proportional to the product of capacitance and frequency. Thus, the higher the frequency, the less the opposition.

Effective (rms) Voltage and Current Relationships: In practice, we usually use effective values rather than peak values. Their ratios are the same, however. Thus reactance can also be expressed as

$$X_C = V_C/I_C \text{ ohms} \qquad (19\text{-}5)$$

where V_C and I_C are the rms values of v_C and i_C respectively. (Note that V_C and I_C are the values that you read on meters, while V_m and I_m are the values that you read on the oscilloscope.)

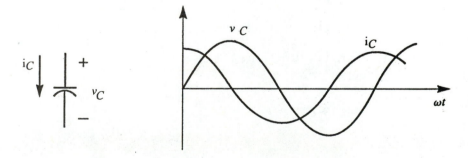

Figure 19-1   Voltage and current for capacitance

MEASUREMENTS

PART A: Phase Relationships for Capacitance

Consider Figure 19-2. Voltage v_S across R_S is in phase with capacitor current i_C. Provided circuit resistance is small compared to X_C, source voltage is approximately equal to the v_C. Thus, the phase difference between Ch1 and Ch2 is approximately equal to the angle between v_C and i_C. It should be close to 90°. Current should lead.

1. a. Assemble the circuit using R_S = 10 Ω and C = 1.0 μF. Set the oscilloscope for dual channel operation.
   b. Set the signal generator to a 500 Hz sine wave. Sketch the waveforms below, appropriately labeled as v_C and i_C. Determine the phase shift between v_C and i_C and indicate on the diagram. How close is it to the theoretical value of 90 degrees?

   c. Vary the frequency a few hundred Hz up and down and observe the phase shift. Is there any appreciable change? Based on this observation, state in your own words the relationship between voltage across and current through capacitance.

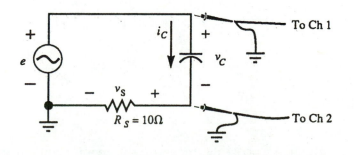

Figure 19-2   Circuit resistance is very small. Therefore, $v_C \approx e$.

PART B: Magnitude Relationships for Capacitance

Preliminary Notes. For the remaining portions of the lab, you need a DMM with a frequency response up to 1 kHz. (Most DMMs can handle this range. However, to be sure, check your meter manual.) If you do not have a suitable meter, make measurements with the oscilloscope.

Since reactance depends on frequency, accurate frequency settings are essential. However, the frequency dial on many signal generators is not accurately calibrated and you may require the oscilloscope to set the frequency. (If you have a generator with a digital readout, the scope should not be necessary.)

2. a. Replace R_S of Figure 19-2 with an accurately measured 220-Ω resistor.

   $R_S =$ _____

   b. Set the signal generator to 500 Hz at 6 V. With a meter (or an oscilloscope if necessary), carefully measure the voltage V_C and the voltage V_S. Calculate current I_C using Ohm's law and the measured value of V_S.

   $V_C =$ _____ $V_S =$ _____ $I_C =$ _____

   c. Calculate the reactance of the capacitor using the measured V_C and I_C.

   $X_{C(measured)} = V_C/I_C =$ _____

   d. If possible, measure C using a bridge. (Otherwise, use the nominal value.) Compute X_C using Equation 19-4. Compare to the measured value of 2(c).

   $C =$ _____ $X_{C(computed)} =$ _____

3. a. Add the second capacitor in series and determine the equivalent reactance of the series combination using the same procedure that you used in Test 2(b) and (c).

   $V_C =$ _____ $V_S =$ _____

   $I_C =$ _____ $X_{eq(measured)} =$ _____

   b. If possible, measure the second capacitor and compute C_{eq} for the series combination. $C_{eq} =$ _____. Use Equation 19-4 to determine $X_{eq(computed)}$. $X_{eq(computed)}$ = _____. Compare to the measured value of 3(a).

4. a. Connect both capacitors in parallel and measure total reactance X_T using the same procedure that you used in Test 2(b) and (c).

V_C = _____ V_S = _____

I_C = _____ $X_{T(measured)}$ = _____

   b. Compute C_T for the parallel combination. C_T = _____.

   Now use Equation 19-4 to determine $X_{T(computed)}$. $X_{T(computed)}$ =

   _____. Compare to the measured value of 4(a).

PART C: Variation of Reactance with Frequency

5. a. Replace the parallel combination with a single 1.0-µF capacitor and measure V_C and V_S and compute reactance (using the procedure of Test 2) at each of the frequencies listed in Table 19-2. Record as $X_{C(measured)}$.
   b. Using Equation 19-4, compute reactance at each of the frequencies and record as $X_{C(computed)}$.

f (Hz)	V_C	V_S	I_C	$X_{C(measured)}$	$X_{C(computed)}$
100					
200					
300					
400					
500					
600					
700					
800					
900					
1000					

Table 19-2

c. Using the graph sheet of Figure 19-3, plot measured and computed reactances versus frequency. Label each plot.

d. Analyze the results. That is, comment on how well the measured values agree with the computed values.

PROBLEMS

6. Consider an ideal capacitor. If you double the capacitance and triple the frequency, what happens to the current if the applied sinusoidal voltage remains the same?

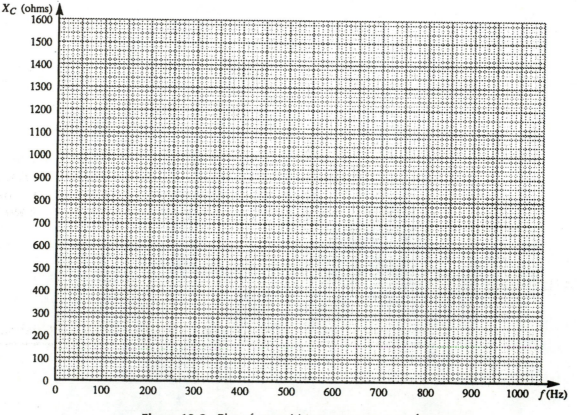

Figure 19-3   Plot of capacitive reactance versus frequency

Power in ac Circuits

OBJECTIVES

After completing this lab, you will be able to
- measure power in a single phase circuit,
- verify power relationships,
- verify power factor relationships,
- determine the effect of adding power factor correction.

EQUIPMENT REQUIRED

☐ Single phase wattmeter
☐ ac ammeter
☐ DMM

COMPONENTS

☐ Resistor: 100-Ω, rated 200-W
☐ Capacitor: 30-μF, non-electrolytic, rated for operation at
              120-VAC (1 required)
☐ Inductor: Approximately 0.2-H, rated to handle 2 amps

Safety Note

In this lab, you will be working with 120 VAC. This voltage is danger-ous, and you must be aware of and observe safety precautions. Famil-iarize yourself with your laboratory's safety features. **Ensure that power is off when you are assembling, changing or otherwise work-ing on your circuit.**

EQUIPMENT USED

Instrument	Manufacturer/Model No.	Serial No.
Single phase wattmeter		
ac Ammeter		

Table 20-1

TEXT REFERENCE

Section 17.7   THE RELATIONSHIP BETWEEN P, Q AND S
Section 17.8   POWER FACTOR
Section 17.9   AC POWER MEASUREMENT

DISCUSSION

Power in ac Systems. Power to an ac load is given by

$$P = V I \cos \theta \qquad (20\text{-}1)$$

where V and I are the magnitudes of the rms load voltage and current respectively, and θ is the angle between them. (θ is the angle of the load impedance.) The quantity $\cos \theta$ is the power factor of the load and is given the symbol F_p. Thus,

$$F_p = \cos \theta \qquad (20\text{-}2)$$

For purely resistive loads, voltage and current are in phase and θ = 0°. Such loads are called *unity power factor* loads. For loads containing resistance and inductance, current lags voltage and their power factor is *lagging*; for loads containing resistance and capacitance, current leads voltage and their power factor is *leading*.

Measuring Power in ac Circuits. Power is measured as in Figure 20-1(a) or (b) with the wattmeter connected so that current passes through its current coil CC and load voltage is applied to its voltage sensing circuit. For this lab, either connection will work.

Power Factor Correction. If a load such as that shown in Figure 20-1 has poor power factor (i.e., θ is large), it will draw excessive current relative to the power transferred to it from the source. One way to improve the power flow in the system is to add power factor correction at the load. Since most power system loads are inductive (because they contain inductive elements such as electric motors, lamp

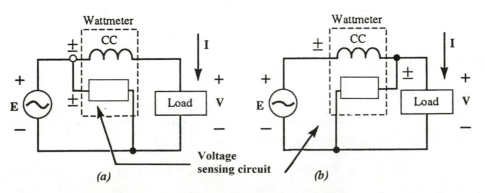

Figure 20-1   Measuring power in an ac circuit

ballasts, and so on), this may be done by adding capacitors in parallel across the load. For loads with a poor power factor, this can dramatically reduce source current. It is important to note, however, that power factor correction does not change the power requirements of the load—it simply supplies the needed reactive power locally, rather than from the source. However, as noted, it greatly reduces source current.

MEASUREMENTS

If possible, use a variable ac power supply (such as a variable autotransformer) as the source—see box.

PART A: Power in a Purely Resistive Circuit

1. Measure the 100-Ω resistor.
   $R =$ _____. Connect the circuit as in Figure 20-2 (next page). (The ammeter may be a standard ac ammeter or the ac current range of a DMM. For the load resistance used here, a 2-A range is adequate. Select a wattmeter to match—e.g., a wattmeter with a 300-W scale.) Carefully measure load voltage, current and power and record in Table 20-2.
2. Using the values of V and R measured in Test 1, compute current, then determine power to the load using each of the formulas $P = VI$, $P = I^2 R$ and $P = V^2/R$. Compare to the value measured with the wattmeter.

> **General Safety Notes**
>
> 1. With power off, assemble your circuit and double check it.
>
> 2. Have your instructor check the circuit before you energize it. If you are using a Powerstat or other variable ac source, gradually increase voltage from zero, watching the meters for signs of trouble.
>
> 3. Turn power off before changing your circuit for the next test.
>
> Check with your instructor for specific safety instructions.

PART B: Power to a Leading Load

3. a. De-energize the circuit and add 30 µF of capacitance in parallel with the load as in Figure 20-3. Measure load voltage, current, and power and record in Table 20-3.

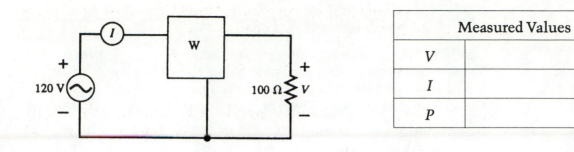

Figure 20-2 Simplified representation of wattmeter.
Use either connection of Figure 20-1 here.

Measured Values	
V	
I	
P	

Table 20-2

    b. Calculate real power using the measured voltage and resistance and compare to the measured value.

    c. Calculate reactive power. $Q =$ _____

    d. Draw the power triangle using the values computed in Tests 3(b) and 3(c). Using the apparent power S from this triangle, compute current. Calculate the percent difference between this and the value measured in Test 3(a). What is the likely source of difference?

    e. Determine the power factor from the data of Table 20-3 and Equation 20-1.

    f. Determine the power factor from the power triangle of 3(d).

g. Determine the load impedance **Z** from Figure 20-3, then calculate power factor using Equation 20-2.

h. Compare the values of power factor determined in Steps (e), (f), and (g)—i.e., comment on any differences and why they might have occurred.

PART C: Power Factor Correction

4. a. Replace the load with an inductive load of approximately 0.2 H as in Figure 20-4. (The exact value is not critical. However, do not exceed the current rating of your ammeter or wattmeter or the power rating of your wattmeter.) Measure load voltage, current, and power and record in Table 20-4.

   b. Add the 30-μF capacitor in parallel with the load as in Figure 20-5 and observe the decrease in current. Explain what is happening. If you add too much capacitance, what will happen?

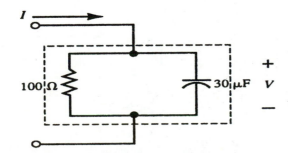

Figure 20-3 Leading power factor load

Measured Values	
V	
I	
P	

Table 20-4

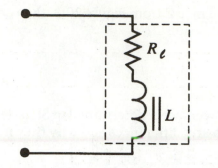

Figure 20-4 Lagging power factor load

Measured Values	
V	
I	
P	

Table 20-4

PROBLEMS

5. A motor delivers 50 hp to a load. The efficiency of the motor is 87% and its power factor is 0.69%. The source voltage is 277 V, 60 Hz.
    a. Determine the power input to the motor.
    b. Draw the power triangle for the motor.
    c. Determine the source current.
    d. Determine how much capacitance is required to correct the power factor to unity.
    e. What is the source current when the capacitance of (d) is added?
    f. If 50% more capacitance than computed in (d) is added, determine the source current.

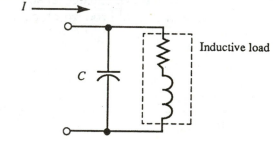

Figure 20-5 Power factor correction

Series ac Circuits

OBJECTIVES

After completing this lab, you will be able to
- calculate current and voltages for a simple series ac circuit,
- measure voltage magnitude and phase angle in a simple series ac circuit,
- verify Kirchhoff's voltage law using measured results,
- measure the internal impedance of a sinusoidal voltage source.

EQUIPMENT REQUIRED

☐ Dual-trace oscilloscope
☐ Signal generator (sinusoidal function generator)
Note: Record this equipment in Table 21-1.

COMPONENTS

☐ Resistors: 680-Ω (1/4-W carbon, 5% tolerance)
☐ Capacitors: 0.22-μF (10% tolerance)

EQUIPMENT USED

Instrument	Manufacturer/Model No.	Serial No.
Oscilloscope		
Signal Generator		

Table 21-1

TEXT REFERENCE

Section 18.1   OHM'S LAW FOR AC CIRCUITS
Section 18.2   AC SERIES CIRCUITS
Section 18.3   KIRCHHOFF'S VOLTAGE LAW &
                THE VOLTAGE DIVIDER RULE

DISCUSSION

Series ac circuits behave in a manner similar to the operation of dc circuits. Ohm's law, Kirchhoff's current and voltage laws and the various circuit analysis rules apply for ac circuits as well as for dc circuits. The principal differences in these rules and laws apply to ac circuits are outlined as follows:

- All impedances are complex numbers and may be expressed either in rectangular form or polar form (e.g., $\mathbf{Z} = 10\,\Omega + j20\,\Omega = 22.36\,\Omega\angle63.43°$). The real component of the impedance vector represents the resistance while the imaginary component corresponds to the reactance. The imaginary component of the impedance will be positive if the reactance is inductive and negative if the reactance is capacitive.
- *Time-domain values* such as $v_C = 2\sin(\omega t + 30°)$ are converted into *phasor domain* ($\mathbf{V}_C = 1.414\,\text{V}\angle30°$) to permit arithmetic operations using complex numbers.
- Although time domain values are always expressed using peak values, phasor voltages and currents are always expressed with magnitudes in rms (root-mean-square).
- Power calculations are performed using rms values and must consider the *power factor* of the circuit or component.

CALCULATIONS

1. Refer to the circuit of Figure 21-1. Determine the reactance of the capacitor at a frequency of $f = 2$ kHz. Express the circuit impedance $\mathbf{Z}_T$ in both the rectangular form and the polar form.

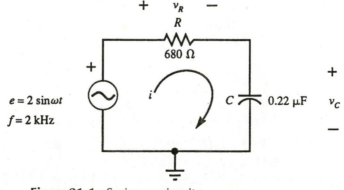

Figure 21-1   Series ac circuit

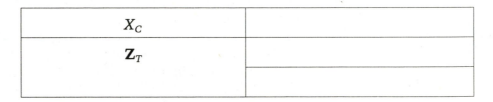

X_C	
$\mathbf{Z}_T$	

2. Convert the time domain form of the voltage source of the circuit of Figure 21-1 into its equivalent phasor domain form. Calculate the phasor current **I** and solve for the phasor voltages $\mathbf{V}_R$ and $\mathbf{V}_C$. Enter your results in Table 21-2.

E	
I	
$\mathbf{V}_R$	
$\mathbf{V}_C$	

Table 21-2

3. Use complex algebra together with the phasor forms of the voltages ($\mathbf{E}$, $\mathbf{V}_R$ and $\mathbf{V}_C$) to verify that Kirchhoff's voltage law applies.

$$\sum \mathbf{V} = 0 = \mathbf{E} - \mathbf{V}_R - \mathbf{V}_C \qquad (21\text{-}1)$$

4. Convert the phasor forms of **I**, $\mathbf{V}_R$, and $\mathbf{V}_C$ into their equivalent time domain values. Enter the results in Table 21-3.

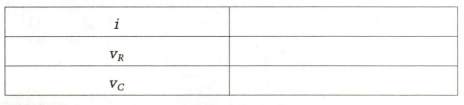

i	
v_R	
v_C	

Table 21-3

MEASUREMENTS

5. Assemble the circuit of Figure 21-1.

6. Refer to the pictorial diagram of Figure 21-2. Connect Ch1 of the oscilloscope to the output of the signal generator. Set the oscilloscope to have an automatic sweep and use Ch1 as the trigger source. Adjust the output of the generator to provide a sinusoidal voltage with an amplitude of 2.0 V (4.0 V_{p-p}) and a frequency of $f = 2$ kHz ($T = 500$ μs).

7. Use Ch2 of the oscilloscope to observe the signal across the capacitor V_C. (You should still have the output of the signal generator displayed on Ch1.) Notice that the voltage across the capacitor is lagging the generator voltage. Adjust the time/division to provide at least one full cycle on the oscilloscope. Accurately sketch and label both the generator voltage e and the capacitor voltage v_C in the space provided on Graph 21-1. In the space below, record the volts/division setting for each channel and the time/division of the oscilloscope.

Ch1 : _____ V/Div.

Ch2 : _____ V/Div.

Time : _____ μs/Div.

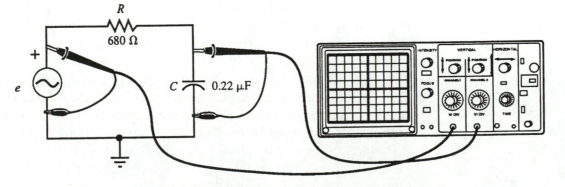

Figure 21-2 Equipment connection for voltage measurement

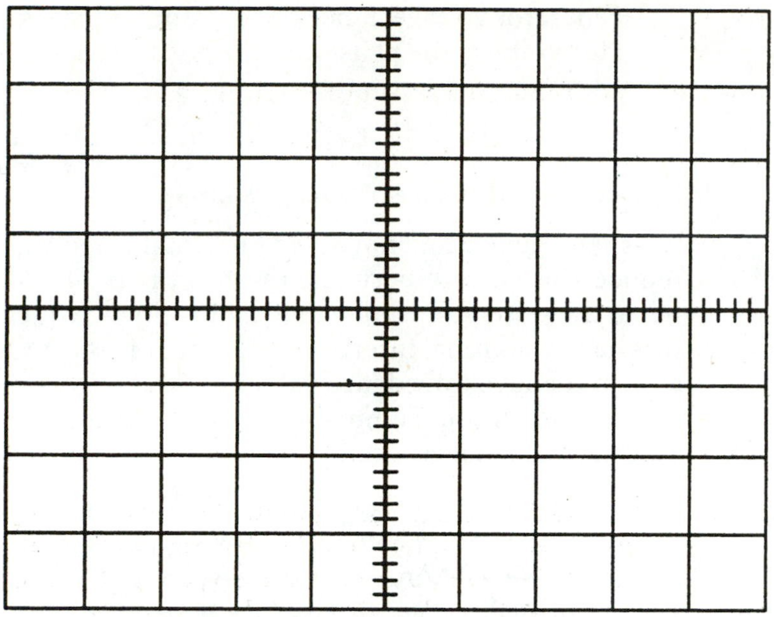

Graph 21-1

8. Use the following expression to calculate the phase angle between capacitor voltage and the generator voltage. (The phase angle is negative since v_C lags **e**.)

$$\theta = \frac{\Delta t}{T} \times 360°$$

(21-2)

θ_1	

9. It is not possible to measure the voltage across the resistor directly since the internal ground connection in the probe of the oscilloscope will result in a short circuit across the capacitor. To overcome this problem, switch the oscilloscope to show the *difference mode* (Ch1 – Ch2), making certain that both Ch1 and Ch2 of the oscilloscope are on the same VOLTS/DIV setting. You should observe a single display, corresponding to the resistor voltage v_R. Since you are using the Ch1 signal as the trigger source, the observed waveform provides the phase shift between the signal generator and the resistor voltage. **Be careful not to adjust the horizontal position. If you accidentally move the display, you will need to readjust the display to obtain the waveforms of Step 7.** You should observe that the resistor voltage is leading the generator voltage. Sketch and label the observed resistor voltage v_R as part of Graph 21-1.

10. Calculate the phase shift between the resistor voltage and the

generator voltage. Since the resistor voltage is leading the generator voltage, the phase angle is positive.

θ_2	

Internal Impedance of Voltage Sources

All signal generators have some internal impedance which tends to reduce the voltage between the terminals of the signal generator when the generator is under load. Most low-frequency signal generators have a nominal internal impedance of 600 Ω. Other signal generators may have impedance values of 50 Ω, 75 Ω or 300 Ω, depending on the application. Figure 21-3 shows a simple representation of the output of a signal generator.

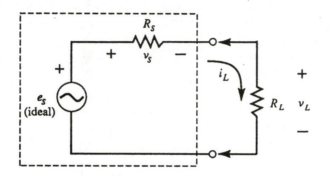

Figure 21-3 Internal impedance of a signal generator

11. Connect the oscilloscope between the output terminals of the signal generator. Adjust the generator to provide a sinusoidal output having an amplitude of 1 V (appearing as 2 $V_{p\text{-}p}$) and a frequency of 1 kHz. Since the signal generator is not loaded, this voltage represents the signal of the ideal voltage source $E_S = 2\ V_{p\text{-}p}$.
12. Connect a 680-Ω load resistor between the output terminals of the signal generator. Reconnect the oscilloscope between the output terminals of the signal generator. You should see that the voltage at the output has now decreased significantly. This represents the loaded output voltage. Measure and record the loaded peak-to-peak output voltage V_L.

V_L	$V_{p\text{-}p}$

CONCLUSIONS

13. Refer to Graph 21-1. Calculate the amplitude V_R of the resistor voltage and the amplitude V_C of the capacitor voltage. From Steps 8 and 10, evaluate the phase angle (with respect to the generator voltage e) for each of these voltages. Express each voltage

in its time domain form [eg. $v_c = V_c \sin(\omega t + \theta_1)$]. Convert the amplitudes into rms quantities and express each voltage in its phasor form. Enter all results in Table 21-4.

Resistor		Capacitor	
V_R	θ	V_C	θ
$v_R =$		$v_C =$	
$\mathbf{V}_R =$		$\mathbf{V}_C =$	

Table 21-4

14. Compare the measured sinusoidal capacitor voltage v_C of Table 21-4 to the theoretical value of Table 21-2.

_____

_____

15. Compare the measured sinusoidal resistor voltage v_R of Table 21-4 to the theoretical value of Table 21-2.

_____

_____

16. Calculate the actual signal generator resistance using Ohm's law and the measurement of Step 12.

R_S	

Parallel ac Circuits

OBJECTIVES

After completing this lab, you will be able to
- measure voltages in a parallel circuit using an oscilloscope,
- use an oscilloscope to indirectly measure current magnitude and phase angles in a simple parallel ac circuit,
- compare measured values to theoretical calculations and verify Kirchhoff's current law,
- determine the power dissipated by a parallel ac circuit.

EQUIPMENT REQUIRED

☐ Dual-trace oscilloscope
☐ Signal generator (sinusoidal function generator)
Note: Record this equipment in Table 22-1.

COMPONENTS

☐ Resistors: 10-Ω (3), 470-Ω (1/4-W carbon, 5% tolerance)
☐ Capacitors: 3300-pF (10% tolerance)
☐ Inductors: 1-mH (iron core, 5% tolerance—Hammond 1534A or equivalent)

EQUIPMENT USED

Instrument	Manufacturer/Model No.	Serial No.
Oscilloscope		
Signal generator		

Table 22-1

TEXT REFERENCE

Section 18.4   AC PARALLEL CIRCUITS
Section 18.5   KIRCHHOFF'S CURRENT LAW & THE CURRENT DIVIDER RULE

DISCUSSION

The equivalent impedance of a parallel circuit is determined by finding the sum of the admittances of all branches. The important point to remember when calculating total admittance is that all admittances are expressed as complex values. This means that we must use complex algebra to find the solution, which is also a complex number.

Refer to the circuit of Figure 22-1. The equivalent admittance of the circuit is determined as

$$\mathbf{Y}_T = \frac{1}{R} + j\frac{1}{X_C} - j\frac{1}{X_L} \tag{22-1}$$

This gives an equivalent circuit impedance of

$$\mathbf{Z}_T = \frac{1}{\mathbf{Y}_T} = \frac{1}{\dfrac{1}{R} + j\dfrac{1}{X_C} - j\dfrac{1}{X_L}} \tag{22-2}$$

In order to further analyze the circuit, it is necessary to convert the ac voltage source into its equivalent phasor form. The circuit current **I** is then easily determined by applying Ohm's law. The current through each component in the circuit is similarly found by applying Ohm's law to each branch or by applying the current divider rule as follows:

$$\mathbf{I}_x = \frac{\mathbf{E}}{\mathbf{Z}_x} = \frac{\mathbf{Z}_T}{\mathbf{Z}_x}\mathbf{I} \tag{22-3}$$

Regardless of the method used to determine currents, Kirchhoff's current law must apply at any node in the circuit. Therefore,

$$\sum \mathbf{I} = 0 \tag{22-4}$$

where each current is in its phasor form.

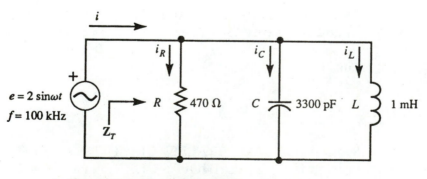

Figure 22-1   Parallel ac circuit

CALCULATIONS

1. Refer to the circuit of Figure 22-1. Determine the reactances of the capacitor and the inductor at a frequency of $f = 100$ kHz. Calculate the circuit impedance $\mathbf{Z}_T$ and express the result in both rectangular and polar form. Enter the data in Table 22-2.

X_C	
X_L	
$\mathbf{Z}_T$	

Table 22-2

2. Convert the time domain form of the voltage source of the circuit of Figure 22-1 into its equivalent phasor domain form. Calculate the phasor current $\mathbf{I}$ and solve for the phasor currents $\mathbf{I}_R$, $\mathbf{I}_C$, and $\mathbf{I}_L$. Enter your results in Table 22-3.

$\mathbf{E}$	
$\mathbf{I}$	
$\mathbf{I}_R$	
$\mathbf{I}_C$	
$\mathbf{I}_L$	

Table 22-3

3. Use complex algebra together with the phasor currents $\mathbf{I}_R$, $\mathbf{I}_C$, and $\mathbf{I}_L$ to verify that Kirchhoff's current law applies at node a.

$$\mathbf{I} = \mathbf{I}_R + \mathbf{I}_C + \mathbf{I}_L \qquad (22\text{-}5)$$

4. Convert the phasor currents $\mathbf{I}$, $\mathbf{I}_R$, $\mathbf{I}_C$, and $\mathbf{I}_L$ into their equivalent time domain forms. Enter the results in Table 22-4.

i	
i_R	
i_C	
i_L	

Table 22-4

Current cannot be measured directly with an oscilloscope. However, by strategically placing small series *sensing resistors* into a circuit and then measuring voltage across these resistors, current through each branch is found by applying Ohm's law.

MEASUREMENTS

5. Assemble the circuit of Figure 22-2. Notice that three 10-Ω *sensing resistors* have been added into the circuit to help in determining branch currents. Since these resistors are small in comparison to the impedance in the branch, they will not significantly load the circuit and their effects may be ignored.
6. Connect Ch1 of the oscilloscope to the output of the signal generator at point a. Set the oscilloscope to have an automatic sweep and use Ch1 as the *trigger source*. Adjust the output of the generator to provide a sinusoidal voltage with an amplitude of 2.0 V (4.0 V_{p-p}) and a frequency of f = 100 kHz (T = 10 μs).
7. Use Ch2 of the oscilloscope to observe the voltage at point b. This is the voltage across sensing resistor R_1. Measure the observed peak-to-peak voltage V_1. Measure the phase angle θ_1 of the voltage V_1 with respect to the generator voltage. Record your results in Table 22-5.

   Since the observed waveform is across the 10-Ω resistor, the amplitude of the current i is now easily calculated from the peak-to-peak voltage as

$$I = \frac{V_1/2}{10\Omega} \qquad (22\text{-}6)$$

   Write the time-domain expression for i using your measurements and calculations. Enter your results in Table 22-5.

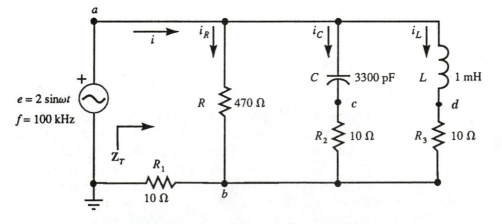

Figure 22-2   Using sensing resistors for current measurement

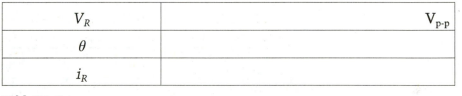

V_1	$V_{p\text{-}p}$
θ_1	
i	

Table 22-5

8. Since we no longer need R_1 in the circuit, it may be removed and replaced with a short circuit. Measure the peak-to-peak voltage V_R and calculate current i_R using Ohm's law. Use your measurements and calculations to record the time-domain expression for i_R in Table 22-6.

V_R	$V_{p\text{-}p}$
θ	
i_R	

Table 22-6

9. Ensure that Ch1 of the oscilloscope is connected to point a (generator voltage) and that the sensing resistor R_1 is removed from the circuit.
   Connect Ch2 of the oscilloscope to point c to measure the voltage across the sensing resistor R_2. Record the peak-to-peak voltage V_2. Measure and record the phase angle between the generator voltage and the sensing voltage V_2. You should observe that the voltage V_2 (and hence, the current) is leading the generator voltage. Use your measurements and Ohm's law to determine the sinusoidal expression for i_c. Record the result in Table 22-7.

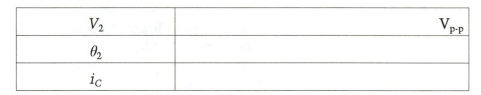

V_2	$V_{p\text{-}p}$
θ_2	
i_C	

Table 22-7

10. With Ch1 of the oscilloscope connected to point a, connect Ch2 of the oscilloscope to point c. Measure and record the peak-to-peak voltage across the sensing resistor R_3. Measure and record the phase angle between the generator voltage and the sensing voltage V_2. You should observe that the voltage V_3 (and hence, the current) is lagging the generator voltage. Use your measurements and Ohm's law to determine the sinusoidal expression for i_L. Record the result in Table 22-8.

V_3		$V_{p\text{-}p}$
θ_3		
i_L		

Table 22-8

CONCLUSIONS

11. Compare the measured currents i, i_R, i_C, and i_L of Table 22-5 through Table 22-8 to the theoretical values of Table 22-4.

    _____

    _____

    _____

12. Convert each of the measured currents i, i_R, i_C, and i_L into its phasor form. Enter the results in Table 22-9.

$\mathbf{I}$	
$\mathbf{I}_R$	
$\mathbf{I}_C$	
$\mathbf{I}_L$	

Table 22-9

13. Use complex algebra to show that the data of Table 22-9 verifies Kirchhoff's current law.

14. Use the measured currents $\mathbf{I}$, $\mathbf{I}_R$, $\mathbf{I}_C$, and $\mathbf{I}_L$ to calculate the total power dissipated by the circuit in Figure 22-2. (You should observe that the sensing resistors dissipate very little power.)

P_T	

Series/Parallel ac Circuits

OBJECTIVES

After completing this lab, you will be able to
- analyze a series/parallel circuit to determine the current through and voltage across each element in a series/parallel circuit,
- measure voltage across each element in a series/parallel circuit using an oscilloscope and use the measurements to determine the current through each element of a series/parallel circuit,
- calculate the power dissipated by each element in a circuit,
- use measurements to verify that the actual powers dissipated correspond to theory.

EQUIPMENT REQUIRED

☐ Dual trace oscilloscope
☐ Signal generator (sinusoidal function generator)
Note: Record this equipment in Table 23-1.

COMPONENTS

☐ Resistors: 10-Ω (2), 470-Ω (1/4-W carbon, 5% tolerance)
☐ Capacitors: 3300-pF (10% tolerance)
☐ Inductors: 1-mH (iron core, 5% tolerance)

EQUIPMENT USED

Instrument	Manufacturer/Model No.	Serial No.
Oscilloscope		
Signal generator		

Table 23-1

TEXT REFERENCE

Section 18.6   SERIES-PARALLEL CIRCUITS

DISCUSSION

The equivalent impedance of a series/parallel ac circuit is determined in a manner which is similar to that used in finding the equivalent resistance of a series/parallel circuit, with the exception that vector algebra is used in determining the total impedance at a given frequency. It is necessary to decide which elements or branches are in series and which are in parallel. The resultant impedance is the combination of the various connections. Once we have the total impedance, it is a simple matter to calculate the total current. By applying appropriate circuit theory, the current, voltage, and power of the various components of the circuit may then be found.

Refer to the circuit of Figure 23-1. Notice that resistor R and inductor L are in parallel. This parallel connection is then seen to be in series with capacitor C. The total impedance of the circuit is therefore determined as

$$\mathbf{Z}_T = -jX_C + R \parallel jX_L \tag{23-1}$$

The power provided to the circuit by the voltage source is calculated as

$$P_T = EI\cos\theta = \frac{E^2}{Z_T}\cos\theta \tag{23-2}$$

In the above expression, E and I are the rms values of the sinusoidal voltage e and current i. Z_T is the magnitude of the circuit impedance and θ is the angle between the current phasor $\mathbf{I} = \mathbf{I}_C$ and the

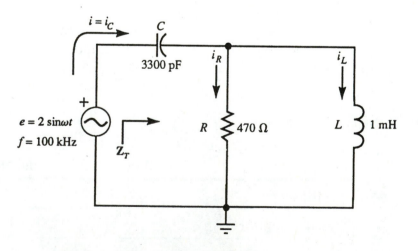

Figure 23-1   Series/parallel circuit

voltage phasor **E**. (This is the same angle as the angle in the impedance vector $\mathbf{Z}_T$.)

On further examination of Figure 23-1, we see that only the resistor can dissipate power. (Inductors and capacitors do not dissipate
power.) This means that the total power in the circuit must also be
the same as the power dissipated by the resistor, namely

$$P_R = \frac{V_R^2}{R} = I_R^2 R$$

(23-3)

where V_R and I_R are rms quantities.

CALCULATIONS

1. Refer to the circuit of Figure 23-1. Determine the reactances of
    the capacitor and the inductor at a frequency of $f = 100$ kHz. Cal-
    culate the circuit impedance $\mathbf{Z}_T$. Enter all values here.

X_C	
X_L	
$\mathbf{Z}_T$	

2. Convert the time-domain form of the voltage source into its
    equivalent phasor-domain form. Calculate the phasor current
    through each element of the circuit. Enter your results in Table
    23-2.

E	
$\mathbf{I}_R$	
$\mathbf{I}_C$	
$\mathbf{I}_L$	

Table 23-2

3. Calculate and record the total power provided to the circuit by the voltage source.

P_T	

4. Use complex algebra to show that currents I_R, I_C, and I_L satisfy Kirchhoff's current law

$$\sum I = 0$$

(23-4)

MEASUREMENTS

5. Assemble the circuit of Figure 23-2. Notice that two 10-Ω *sensing resistors* have been added to the circuit to help in determining branch currents. Since these resistors are small in comparison to the impedance in the branch, they will not significantly load the circuit.

6. Connect Ch1 of the oscilloscope to the output of the signal generator at point a. Set the oscilloscope to an automatic sweep and use Ch1 as the *trigger source*. Adjust the output of the generator to provide a sinusoidal voltage with an amplitude of 2.0 V (4.0 V_{p-p}) at a frequency of $f = 100$ kHz ($T = 10$ μs).

7. Use Ch2 of the oscilloscope to observe the voltage at point c. This is the voltage across sensing resistor R_1. Measure the peak-to-peak voltage V_1. Determine the phase angle θ_1 of voltage v_1 with respect to the generator voltage e. Write the time-domain expression for i using your measurements and calculations. Record your results in Table 23-3.

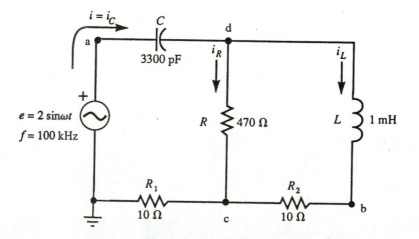

Figure 23-2   Using sensing resistors for current measurement

V_1		$V_{p\text{-}p}$
θ_1		
$i = i_c$		

Table 23-3

8. Remove resistor R_1 from the circuit and replace it with a short circuit. Move Ch2 of the oscilloscope to observe the voltage at point d (the voltage across resistor R). Measure the peak-to-peak voltage V_R and determine phase angle θ_R of voltage v_R with respect to the generator voltage e. Record your results in Table 23-4. Determine the amplitude i_R and enter the time-domain expression for i_R in Table 23-4.

V_R		$V_{p\text{-}p}$
θ_R		
i_R		

Table 23-4

9. With Ch1 of the oscilloscope connected to point a, connect Ch2 of the oscilloscope to point b. Measure and record the peak-to-peak voltage across the sensing resistor R_2. Measure and record the phase angle between the generator voltage and the sensing voltage V_2. Use your measurements and Ohm's law to determine the sinusoidal expression for i_L. Record the result in Table 23-5.

V_2		$V_{p\text{-}p}$
θ_2		
i_L		

Table 23-5

CONCLUSIONS

10. Convert the measured currents i_R, i_C, and i_L of Table 23-3 through Table 23-5 into their phasor forms. Enter your results in Table 23-6 and compare them with the theoretical values of Table 23-2.

$I = I_C$	
I_R	
I_L	

Table 23-6

_____

_____

11. Use complex algebra to verify that the data of Table 23-6 satisfy Kirchhoff's current law.

12. Use the phasors **E** and **I** to calculate the total power delivered to the circuit by the voltage source. Enter your calculation in the space provided.

P_T	

13. Now use the rms values of V_R and I_R to determine the power dissipated by the resistor. Enter the result in the space below. How does this value compare with the total power delivered to the circuit by the voltage source? Offer an explanation for any variation.

P_R	

_____

_____

PROBLEMS

14. Refer to the circuit of Figure 23-1. If the frequency of the signal generator is increased to 200 kHz, determine the following:
    a. Total impedance Z_T,
    b. Currents **I**, I_R, and I_L,
    c. Power P_R dissipated by the resistor,
    d. Power P_T delivered by the voltage source.
15. Repeat Problem 14 if the frequency of the generator is decreased to 50 kHz.

Thévenin's and Norton's Theorems

OBJECTIVES

After completing this lab, you will be able to

- calculate the Thévenin and Norton equivalents of an ac circuit,
- measure the Thévenin (open circuit) voltage and the Norton (short circuit) current of an ac circuit,
- calculate the Thévenin impedance of a circuit using the measured values of Thévenin voltage and Norton current,
- measure the load impedance which results in a maximum transfer of power to the load.

EQUIPMENT REQUIRED

☐ Dual trace oscilloscope
☐ Signal generator (sinusoidal function generator)
☐ DMM
 Note: Record this equipment in Table 24-1.

COMPONENTS

☐ Resistors: 10-Ω, 1.5-kΩ (1/4-W carbon, 5% tolerance)
              5-kΩ variable resistor
☐ Capacitors: 2200-pF (10% tolerance)
☐ Inductors: 2.4-mH (iron core, 5% tolerance)

EQUIPMENT USED

Instrument	Manufacturer/Model No.	Serial No.
Oscilloscope		
Signal generator		
DMM		

Table 24-1

TEXT REFERENCE

DISCUSSION

Thévenin's theorem allows us to convert any two-terminal linear bilateral circuit into an equivalent circuit consisting of a voltage source E_{Th} in series with an impedance Z_{Th} as illustrated in Figure 24-1(a). When a load is connected across the two terminals of the circuit, the current through (or voltage across) the load is easily calculated by analyzing the equivalent circuit. The Thévenin voltage of an equivalent circuit is determined by removing the load from the circuit and measuring the open circuit voltage.

Norton's theorem is the duality of Thévenin's theorem in that it converts any two-terminal linear bilateral circuit into an equivalent circuit consisting of a current source I_N in parallel with an impedance Z_N as shown in Figure 24-1(b). The Norton current of an equivalent circuit is determined by replacing the load with a short circuit and measuring the current through the load.

Unlike dc circuits, the Thévenin (and Norton) impedance of an ac circuit cannot be measured directly. Rather, the impedance is determined indirectly by using the equalence between the Thévenin and Norton circuits. Since the circuits are equivalent, the following relationship must apply:

$$Z_{Th} = Z_N = \frac{E_{Th}}{I_N} \tag{24-1}$$

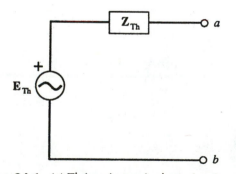

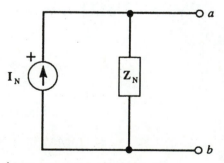

Figure 24-1   (a) Thévenin equivalent circuit   (b) Norton equivalent circuit

CALCULATIONS

1. Refer to the circuit of Figure 24-2. Determine the reactances of the capacitor and the inductor at a frequency of $f = 100$ kHz. Enter the results here.

X_C	
X_L	

2. Convert the time-domain form of the voltage source in Figure 24-2 into its equivalent phasor-domain form and enter the result here.

E	

3. Determine the Thévenin equivalent circuit to the left of terminals a and b in the circuit of Figure 24-2. Sketch the equivalent circuit in this space.

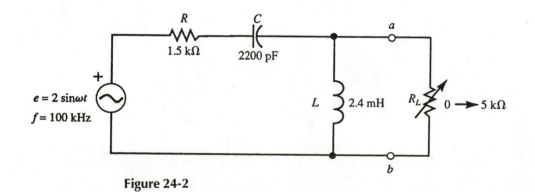

Figure 24-2

4. Determine the Norton equivalent circuit to the left of terminals a and b in the circuit of Figure 24-2. Sketch the equivalent circuit in this space.

5. *Absolute maximum power* will be delivered to the load imped-ance when the load is the complex conjugate of the Thévenin (or Norton) impedance. For what value of load impedance $\mathbf{Z}_L$ will the circuit of Figure 24-2 transfer maximum power to the load? Solve for the absolute maximum load power.

$\mathbf{Z}_L$	
P_{max}	

6. The load in the circuit of Figure 24-2 does not have any reactance. In this case, absolute maximum power cannot be transferred to the load. *Relative maximum power* will be transferred to the load impedance when the value of load resistance is equal to the fol-lowing:

$$R = \sqrt{R_{Th}^2 + X_{Th}^2}$$

(24-2)

For what value of load resistance R_L will the circuit of Figure 24-2 transfer maximum power to the load? Solve for the relative maxi-mum load power and enter the results in the space provided.

R_L	
P_{max}	

MEASUREMENTS

7. Assemble the circuit of Figure 24-2, temporarily omitting the load resistor R_L. Connect Ch1 of the oscilloscope to the output of the signal generator and adjust the generator to provide an output of 2.0 V_p (4.0 V_{p-p}) at $f = 100$ kHz.

8. While using Ch1 as the trigger source of the oscilloscope, connect Ch2 between terminals a and b of the circuit. Measure and record the peak-to-peak value of the open circuit voltage amplitude E_{Th} and the phase shift θ. (The phase shift is measured with respect to the generator voltage observed on Ch1.)

E_{Th}	V_{p-p}
θ	

9. Place a 10-Ω *sensing resistor* between terminals a and b. The sensing resistor has a very low impedance with respect to the other components in the circuit and so has a minimal loading effect. Readjust the generator voltage to ensure that the output is 2.0 V_p. Measure the peak-to-peak voltage across the sensing resistor and determine the peak-to-peak value of the "short circuit" current I_N and the phase shift θ.

V_{ab}	V_{p-p}
I_N	mA_{p-p}
θ	

10. Use the DMM to adjust the 5-kΩ variable resistor for a resistance of 500 Ω. Remove the sensing resistor and insert the variable resistor between terminals a and b of the circuit. Adjust the supply voltage for 2.0 V_{p-p} and measure the amplitude of the output voltage, V_L. Enter your measurement in Table 24-2.

11. Remove R_L from the circuit and incrementally increase the resistance by 500 Ω. Repeat Step 10. Keep increasing the load resistance until $R_L = 5$ kΩ. Enter all data in Table 24-2.

R_L	V_L
500 Ω	
1000 Ω	
1500 Ω	
2000 Ω	
2500 Ω	
3000 Ω	
3500 Ω	
4000 Ω	
4500 Ω	
5000 Ω	

Table 24-2

CONCLUSIONS

12. Convert the measured voltage and phase angle for the Thévenin voltage of Step 8 into its correct phasor form. Record your result in the space provided below. How does this value compare to the theoretical value determined in Step 3?

E_{Th}	

_____

_____

13. Convert the measured current and phase angle for the Norton current of Step 9 into its correct phasor form. Enter the result below. How does this value compare to the theoretical value determined in Step 4?

I_N	

_____

_____

14. Use the phasors of Steps 12 and 13 to calculate the Thévenin

(and Norton) impedance. How does this value compare to the theoretical value determined in Steps 3 and 4?

$Z_{Th} = Z_N$	

15. Use the data of Table 24-2 to calculate the power delivered to the load for each of the resistor values. (Remember that you will need to convert each voltage measurement into its equivalent rms value in order to calculate the power.) Enter the results in Table 24-3.

R_L	V_L
500 Ω	
1000 Ω	
1500 Ω	
2000 Ω	
2500 Ω	
3000 Ω	
3500 Ω	
4000 Ω	
4500 Ω	
5000 Ω	

Table 24-3

16. Use the data of Table 24-2 to sketch a graph of power (in microwatts) versus load resistance (in ohms). Connect the points with the best smooth continuous curve. (A correctly drawn curve will not be drawn from point-to-point.)

17. Use the curve of Graph 24-1 to determine the approximate value of load resistance R_L for which the load receives maximum power from the circuit. Enter the result here. How does this value compare to the value determined in Step 6?

R_L	

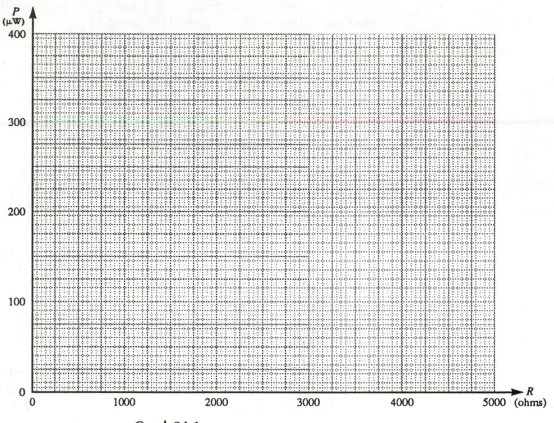

Graph 24-1

PROBLEMS

18. If the circuit of Figure 24-2 (f = 100 kHz) is to provide absolute maximum power to the load, what value of capacitance (in µF or inductance (in mH) must be added in series with the load resistance?

19. Determine the Norton equivalent of the circuit to the left of points a and b in the circuit of Figure 24-2. Assume that the circuit operates at f = 200 kHz.

20. Refer to the circuit of Figure 24-2.
    a. Determine the Thévenin equivalent of the circuit to the left of points a and b, assuming that the circuit operates at a frequency of f = 50 kHz.
    b. Solve for the load impedance which will result in a maximum transfer of power to the load.
    c. Calculate the maximum power which can be transferred to the load.

LAB 25

Series Resonance

OBJECTIVES

After completing this lab, you will be able to
- calculate the resonant frequency of a series resonant circuit,
- solve for the maximum output voltage of a resonant circuit using the *quality factor Q* of the circuit,
- measure the bandwidth of a series resonant circuit,
- measure the impedance at frequencies above and below the resonant frequency and observe that it is purely resistive only at resonance,
- sketch the circuit current as a function of frequency and explain why the response has a bell-shaped curve when plotted on a semi-logarithmic graph.

EQUIPMENT REQUIRED

☐ Dual trace oscilloscope
☐ Signal generator (sinusoidal function generator)
☐ DMM
Note: Record this equipment in Table 25-1.

COMPONENTS

☐ Resistors: 15-Ω (1/4-W carbon, 5% tolerance)
☐ Capacitors: 0.33-μF (10% tolerance)
☐ Inductors: 2.4-mH (iron core, 5% tolerance)

EQUIPMENT USED

Instrument	Manufacturer/Model No.	Serial No.
Oscilloscope		
Signal generator		
DMM		

Table 25-1

TEXT REFERENCE

DISCUSSION

Resonant circuits are used throughout electronics as a means of passing a range of frequencies, while rejecting all other frequencies. These circuits have important applications in communications where they are used in circuits such as receivers to tune into a particular station or channel. Figure 25-1 represents a typical series resonant circuit.

At the resonant frequency, the reactance of the inductor is exactly equal to the reactance of the capacitor. Since they are equal with opposite phase, the reactances cancel and the total impedance of the circuit is purely resistive at resonance. The resonant frequency (in hertz) of a series circuit is given as

$$f_S = \frac{1}{2\pi\sqrt{LC}} \qquad (25\text{-}1)$$

At the resonant frequency, the current (and power) in the circuit is maximum, resulting in a maximum output voltage appearing across the inductor. Since the reactance of the inductor can be many times greater than the resistance of the circuit, the output voltage may be many times greater than the applied signal. This characteristic is one of the advantages of using a resonant circuit since the out-

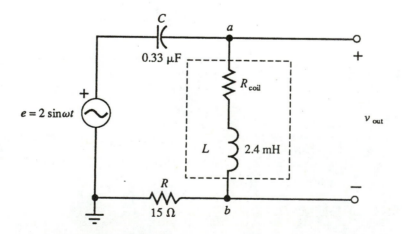

Figure 25-1   Series resonant circuit

put voltage is amplified without the need for *active components* such as transistors.

The *quality factor* Q of a resonant circuit is defined as the ratio of reactive power to the real power at the resonant frequency. It can be shown that the quality factor for the circuit of Figure 25-1 is determined as

$$Q = \frac{X_L}{R} = \frac{\theta L}{R_S + R_{coil}}$$ (25-2)

The Q of a circuit is used to determine the range of frequencies which will be passed by a given resonant circuit. If the circuit has a high Q (greater than 10), it will pass a narrow range of frequencies and the circuit is said to have a *high selectivity*. Conversely, if the Q of the circuit is small, the circuit will pass a broader range of frequencies and the circuit is said to have a *low selectivity*. The *bandwidth (BW)* of a resonant circuit is defined as the difference between the the half-power frequencies, namely the frequencies at which the circuit dissipates half the power that would be dissipated at resonance. The bandwidth (in hertz) of a resonant circuit is determined as

$$BW = \frac{f_S}{Q}$$ (25-3)

The half-power frequencies occur on either side of the resonant frequency. If the quality factor of the circuit is large (Q ≥ 10) then the half-power frequencies are given approximately as

$$f_1 \cong f_S - \frac{BW}{2}$$ (25-4)

and

$$f_2 \cong f_S + \frac{BW}{2}$$ (25-5)

CALCULATIONS

1. Determine the resonant frequency for the circuit of Figure 25-1.

f_S	

2. Prior to starting the lab, obtain an inductor from your lab instructor. Use the DMM ohmmeter to measure the dc resistance of the inductor. Enter the result below.

R_{coil}	

3. Using the measured resistance of the inductor, calculate the phasor form of current **I** at resonance. Enter the result in the space provided below.

I	

4. Determine the phasor form of output voltage $\mathbf{V}_{out}$ appearing across the inductor. You will need to consider the effect of R_{coil}. Enter the result below.

$\mathbf{V}_{out}$	

5. Calculate the quality factor Q, bandwidth BW, and approximate half-power frequencies f_1 and f_2 for the circuit. Record the results in the space provided.

Q	
BW	
f_1	
f_2	

MEASUREMENTS

6. Assemble the circuit shown in Figure 25-1. Connect Ch1 of the oscilloscope to the output of the signal generator and adjust the generator to provide an amplitude of 2.0 V_p (4.0 V_{p-p}) at a frequency of $f = 1$ kHz.

7. Connect Ch2 of the oscilloscope across the resistor R and measure the amplitude of V_R. (In order to simplify this measurement, it is generally easier to measure the peak-to-peak voltage and then divide by two.) Use the measured voltage to calculate the amplitude of the current I. Enter the results in Table 25-2 (next page).

8. Increase the frequency of the signal generator to the frequencies indicated in Table 25-2. Due to loading effects, the amplitude of the generator will tend to drift. Ensure that the output of the signal generator is kept constant at 2.0 V_p. Measure the amplitude of V_R for each frequency and calculate the corresponding amplitude of current I. Enter the values in Table 25-2.

f	V_R	I
1 kHz		
2 kHz		
3 kHz		
4 kHz		
4.5 kHz		
5 kHz		
5.5 kHz		
6 kHz		
6.5 kHz		
7 kHz		
8 kHz		
9 kHz		
10 kHz		

Table 25-2

9. For which frequency f in Table 25-2 is the circuit current a maximum? Adjust the generator to provide an output of 2.0 V_p at this frequency. While observing the oscilloscope, adjust the the generator frequency until the resistor voltage V_R is at the maximum value. Record the resonant frequency f_s and the corresponding resistor voltage V_R in Table 25-3. Calculate the current at resonance.

10. Decrease the frequency until the output voltage V_R is reduced to 0.707 of the maximum value found in Step 9. Record the lower half-power frequency f_1 and the corresponding resistor voltage V_R in Table 25-3. (Ensure that the output of the signal generator is at 2.0 V_p.)

    Increase the frequency above the resonant frequency until the output voltage V_R is again reduced to 0.707 of the maximum value found in Step 9. Record the upper half-power frequency f_2 and the corresponding resistor voltage V_R in Table 25-3. (Ensure that the output of the signal generator is at 2.0 V_p.)

    Calculate the current for each frequency.

f	V_R	I
$f_1 =$		
$f_S =$		
$f_2 =$		

Table 25-3

11. Set the signal generator to the resonant frequency determined in Step 9 and adjust the amplitude for 2.0 V_p. With Ch1 of the oscilloscope at the output of the signal generator, use Ch2 to measure the voltage across resistor R. Calculate the phasor form of current **I** at resonance. (You should observe that v_R and e are in phase.) Record your results here.

V_R	
θ	
I	

12. Adjust the signal generator for a frequency $f = f_1$ and a voltage of 2.0 V_p. Measure the magnitude and phase angle of the voltage V_R. Calculate the phasor form of current **I** at this frequency.

V_R	
θ_1	
I_1	

13. Adjust the signal generator for a frequency $f = f_2$ and a voltage of 2.0 V_p. Measure the magnitude and phase angle of the voltage V_R. Calculate the phasor form of the current **I** at this frequency.

V_R	
θ_2	
I_2	

14. Set the signal generator to the resonant frequency determined in Step 9 and adjust the amplitude for 2.0 V_p. Place Ch1 of the oscilloscope at terminal a of the circuit and Ch2 at terminal b. Use the difference mode of the oscilloscope to measure the amplitude of the output voltage. You should observe that the amplitude of this voltage is larger than the amplitude at the output of the signal generator. Record the amplitude of output voltage V_{out} in the space provided.

V_{out}	

CONCLUSIONS

15. Plot the data of Tables 25-2 and 25-3 on the semi-logarithmic scale of Graph 25-1. Connect the points with the best smooth continuous curve.

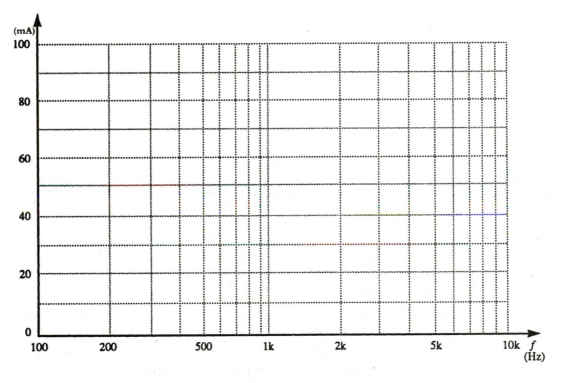

Graph 25-1

16. Compare the measured resonant frequency f_S found in Step 9 to the theoretical frequency determined in Step 1.

_____

_____

17. Compare the measured half-power frequencies from Step 10 to the theoretical values determined in Step 5.

_____

_____

_____

_____

18. Use the measured half-power frequencies to calculate the bandwidth of the circuit. Enter the results here.

$$BW = f_2 - f_1 \qquad\qquad (25\text{-}6)$$

BW	

19. Calculate the Q of the circuit and enter the result in the space provided.

$$Q = \frac{f_S}{BW} \qquad\qquad (25\text{-}7)$$

Q	

20. Use the data of Step 11 to compare the phase angle of the current with respect to the signal generator at resonance. Based on this result, is the circuit resistive, inductive, or capacitive when $f = f_S$?

_____

_____

21. Use the data of Step 12 to compare the phase angle of the current with respect to the signal generator when $f = f_1$. Based on this result, is the circuit resistive, inductive, or capacitive when $f < f_S$?

_____

_____

22. Use the data of Step 13 to compare the phase angle of the current with respect to the signal generator when $f = f_2$. Based on this result, is the circuit resistive, inductive, or capacitive when $f > f_S$?

_____

_____

23. In Step 14 , you should have observed that the output voltage of the circuit at resonance is larger than the applied signal generator voltage. Calculate and record the ratio of the amplitudes V_{out}/E. Since the output voltage is taken across the inductor, you should observe that this ratio is very close to the expected Q of the circuit. Compare this result with that obtained in Step 19.

$Q = V_{out}/E$	

PROBLEMS

24. If the resistance of the inductor R_{coil} was higher than the measured value, what would happen to f_S, Q, BW, and the output voltage v_{out} at resonance?

f_S:

Q:

BW:

---

v_{out} at resonance:

Parallel Resonance

OBJECTIVES

After completing this lab, you will be able to
- calculate the resonant frequency of a parallel resonant circuit,
- solve for the maximum output voltage of a parallel resonant circuit,
- measure the bandwidth of a parallel resonant circuit,
- measure the impedance at frequencies above and below the resonant frequency and observe that it is purely resistive at resonance,
- sketch the output voltage as a function of frequency and explain why the response has a bell-shaped curve when plotted on a semi-logarithmic graph.

EQUIPMENT REQUIRED

☐ Dual trace oscilloscope
☐ Signal generator (sinusoidal function generator)
☐ DMM
Note: Record this equipment in Table 26-1.

COMPONENTS

☐ Resistors: 1-kΩ, 1.5-kΩ (1/4-W carbon, 5% tolerance)
☐ Capacitors: 0.33-μF (10% tolerance)
☐ Inductors: 2.4-mH (iron core, 5% tolerance)

EQUIPMENT USED

Instrument	Manufacturer/Model No.	Serial No.
Oscilloscope		
Signal generator		
DMM		

Table 26-1

TEXT REFERENCE:

Section 21.2   QUALITY FACTOR, Q
Section 21.6   PARALLEL RESONANCE

DISCUSSION

Although series resonant networks are used occasionally in electrical and electronic circuits, parallel resonant networks are the most common type used. Figure 26-1 illustrates a simple parallel resonant network, often called an *LC tank circuit*.

The impedance of the tank circuit is relatively low at all frequencies except at the frequency of resonance. At the resonant frequency, the reactance of the capacitor is exactly equal to the reactance of the inductor. The resulting parallel impedance approaches that of an open circuit. Since the inductor will always have some series resistance due to the coil of wire, the actual impedance of the tank circuit will not be infinitely large. In practice, the impedance of the tank circuit will generally have an impedance between 10 kΩ and 100 kΩ. When the tank circuit is connected across a constant current source (usually a transistor), the voltage across the tank circuit will be relatively high at the resonant frequency and very low at all other frequencies.

The resonant frequency of a tank circuit is found to be

$$f_P = \frac{1}{2\pi\sqrt{LC}}\sqrt{1 - \frac{R_{\text{coil}}^2 C}{L}} \qquad (26\text{-}1)$$

If R_{coil}^2 is at least ten times smaller than the ratio L/C, then the parallel resonant frequency may be approximated as

$$f_P = \frac{1}{2\pi\sqrt{LC}} \qquad (26\text{-}2)$$

The input impedance of a tank circuit at resonance will always be purely resistive, and may be determined by using the *quality factor* Q of the coil as follows:

$$R_P = (Q_{\text{coil}}^2 + 1)R_{\text{coil}} \qquad (26\text{-}3)$$

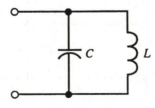

Figure 26-1   Ideal LC tank circuit

If a resistance R_1 is placed in parallel with the tank, the Q of the circuit will be reduced since this resistor absorbs some of the energy from the circuit. Also, if the voltage source has a series resistance R_s, then the Q of the circuit is reduced still further. For the network shown, the quality factor is determined as

$$Q = \frac{R_{eq}}{X_c} \quad \text{where } R_{eq} = R_1 \| R_P \| R_S \qquad (26\text{-}4)$$

Notice that the Q of the circuit is determined by placing R_s in parallel with the other resistors. The reason for this becomes apparent if the volage source and its series resistance are converted into an equivalent current source and parallel resistance.

As in the series resonant circuit, the quality factor may be used to determine the bandwidth of the circuit as

$$BW = \frac{f_P}{Q} \qquad (26\text{-}5)$$

CALCULATIONS

1. Calculate and record the resonant frequency for the circuit of Figure 26-2.

2. Prior to starting the lab, obtain an inductor from your lab instructor. Use the DMM ohmmeter to measure the dc resistance of the inductor. Enter the result here.

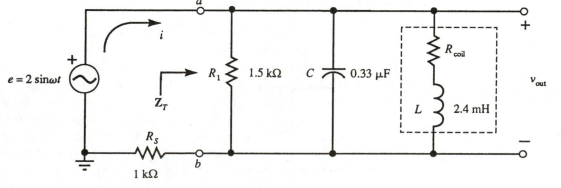

Figure 26-2 Parallel resonant circuit

3. Calculate the equivalent impedance R_P of the LC tank at resonance. Use this value to determine the total impedance Z_T of the circuit at resonance. (The impedance will be resistive.) Calculate the phasor form of current $\mathbf{I}$ at resonance and determine the output voltage phasor $\mathbf{V}_{out}$. Enter your results below.

R_P	
$\mathbf{Z}_T$	
$\mathbf{I}$	
$\mathbf{V}_{out}$	

4. Calculate and record the quality factor Q, bandwidth BW, and half-power frequencies f_1 and f_2 for the circuit.

Q	
BW	
f_1	
f_2	

MEASUREMENTS

5. Assemble the circuit shown in Figure 26-2. Connect Ch1 of the oscilloscope to the output of the signal generator and adjust the output to have an amplitude of 2.0 V_p (4.0 V_{p-p}) at a frequency of $f = 1$ kHz.
6. Use Ch1 as the trigger source and connect Ch2 of the oscilloscope across the resistor R_S. Set the oscilloscope on the difference mode to display the amplitude of the sinusoidal output voltage. Measure the amplitude of voltage V_{out} and record the results in Table 26-2.
7. Increase the frequency of the signal generator to the frequencies indicated in Table 26-2. Adjust the amplitude of the signal generator to maintain an amplitude of 2.0 V_p for each frequency. Measure and record the amplitude of the output voltage for each frequency.

f	V_{out}
1 kHz	
2 kHz	
3 kHz	
4 kHz	
4.5 kHz	
5 kHz	
5.5 kHz	
6 kHz	
6.5 kHz	
7 kHz	
8 kHz	
9 kHz	
10 kHz	

Table 26-2

8. For which frequency f in Table 26-2 is the output voltage a maximum? Adjust the generator to provide an output of 2.0 V_p at this frequency. While observing the oscilloscope, adjust the the generator frequency until voltage V_{out} is at the maximum value. Record the resonant frequency f_P and the corresponding output voltage V_{out} in Table 26-3.

9. Decrease the frequency until the output voltage V_{out} is reduced to 0.707 of the maximum value found in Step 8. Record the lower half-power frequency f_1 and the corresponding output voltage V_{out} in Table 26-3. (Ensure that the output of the signal generator is at 2.0 V_p.)

   Increase the frequency above the resonant frequency until the output voltage V_{out} is again reduced to 0.707 of the maximum value found in Step 8. Record the upper half-power frequency f_2 and the corresponding resistor voltage V_{out} in Table 26-3. (Ensure that the output of the signal generator is at 2.0 V_p.)

f	V_{out}
$f_1 =$	
$f_P =$	
$f_2 =$	

Table 26-3

10. Set the signal generator to the resonant frequency determined in Step 9 and adjust the amplitude for 2.0 V_p. Measure the magnitude and phase angle (with respect to the signal generator) of the voltage across R_S. Calculate the phasor form of current **I** at resonance. (You should observe that v_S and e are in phase.) Record your results.

V_S	
θ	
I	

11. Adjust the signal generator for a frequency $f = f_1$ and a voltage of 2.0 V_p. Measure the magnitude and phase angle of the voltage V_S. Calculate the phasor form of current **I** at this frequency. Record these values.

V_R	
θ	
I	

12. Adjust the signal generator for a frequency $f = f_2$ and a voltage of 2.0 V_p. Measure the magnitude and phase angle of the voltage V_S. Calculate the phasor form of the current **I** at this frequency. Record the values.

V_S	
θ	
I	

CONCLUSIONS

13. Plot the data of Tables 26-2 and 26-3 on the semi-logarithmic scale of Graph 26-1. Connect the points with the best smooth continuous curve.

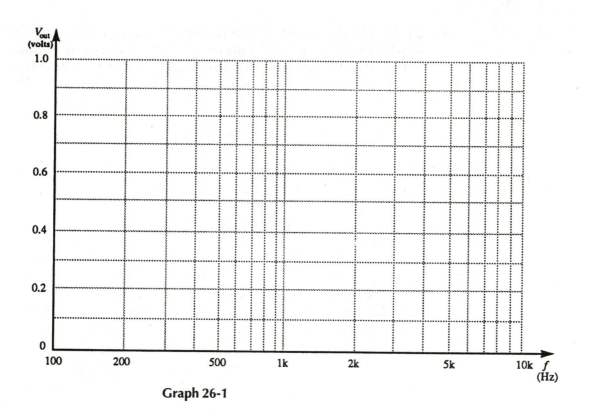

Graph 26-1

14. Compare the measured resonant frequency f_P as found in Step 8 to the theoretical frequency determined in Step 1.

_____

_____

15. Compare the measured half-power frequencies of Step 9 to the theoretical values determined in Step 3.

_____

_____

16. Calculate and record the bandwidth.

$$BW = f_2 - f_1 \qquad (26\text{-}6)$$

BW	

17. Calculate and record the quality factor Q of the circuit.

$$Q = \frac{f_S}{BW} \qquad (26\text{-}7)$$

Q	

Compare this value of Q to that calculated in Step 3.

_____

_____

18. Use the data of Step 10 to compare the phase angle of the current with respect to the signal generator at resonance. Based on this result, is the circuit resistive, inductive, or capacitive when $f = f_P$?

_____

_____

19. Use the data of Step 11 to compare the phase angle of the current with respect to the signal generator when $f = f_1$. Based on this result, is the circuit resistive, inductive, or capacitive when $f < f_P$?

_____

_____

20. Use the data of Step 12 to compare the phase angle of the current with respect to the signal generator when $f = f_2$. Based on this result, is the circuit resistive, inductive, or capacitive when $f > f_P$?

_____

_____

PROBLEMS

21. If the resistance of the inductor R_{coil} was higher than the measured value, what would happen to R_P, Q, and BW at resonance?
    R_P:

_____

Q:

_____

BW:

_____

LAB 27

RC and *RL* Low-Pass Filter Circuits

OBJECTIVES

After completing this lab, you will be able to
- develop the transfer function for a low-pass filter circuit,
- determine the cutoff frequency of a low-pass filter circuit,
- sketch the Bode plot of the transfer function for a low-pass filter,
- compare the measured voltage gain response of a low-pass filter to the theoretical asymptotic response predicted by a Bode plot,
- explain why the voltage gain of a low-pass filter drops at a rate of 20 dB for each decade increase in frequency.

EQUIPMENT REQUIRED

☐ Dual trace oscilloscope
☐ Signal generator (sinusoidal function generator)
☐ DMM
Note: Record this equipment in Table 27-1.

COMPONENTS

☐ Resistors: 75-Ω, 330-Ω (1/4-W carbon, 5% tolerance)
☐ Capacitors: 0.47-μF (10% tolerance)
☐ Inductors: 2.4-mH (iron core, 5% tolerance)

EQUIPMENT USED

Instrument	Manufacturer/Model No.	Serial No.
Oscilloscope		
Signal generator		

Table 27-1

TEXT REFERENCE

Section 22.3   SIMPLE *RC* AND *RL* TRANSFER FUNCTIONS
Section 22.4   THE LOW-PASS FILTER CIRCUIT

DISCUSSION

Filter circuits are used extensively in electrical and electronic circuits to remove unwanted signals while permitting desired signals to pass from one stage to another. Although there are many types of filter circuits, most filters are *low-pass, high-pass, band-pass,* or *band-reject* filters. As the name implies, *low-pass* filters permit low frequencies to pass from one stage to another. Figure 27-1 shows both *RC* and *RL* low-pass filters.

The frequency at which the low-pass filter begins to attenuate (decrease the amplitude of) a signal is called the *cutoff frequency* or *break frequency* f_C. Specifically, the cutoff frequency is that frequency at which the amplitude of the output voltage is 0.707 of the amplitude of low frequency signals. Since this frequency corresponds to half power, the cutoff frequency is also called the *half-power* or *3-dB down frequency*. (Recall that when the output power is half of the input power, the attenuation of the stage is 3 dB.)

Cutoff frequencies are generally calculated as ω_C in radians per second, since the algebra tends to be fairly straightforward. However, when working with measurements it is easiest to use frequencies expressed as f_C in hertz. The cutoff frequencies for an *RC* low-pass filter are given as follows:

$$\omega_C = \frac{1}{\tau} = \frac{1}{RC} \tag{27-1}$$

and

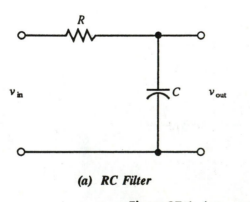

(a) RC Filter

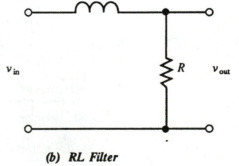

(b) RL Filter

Figure 27-1   Low-pass filters

$$f_C = \frac{\omega_C}{2\pi} = \frac{1}{2\pi RC} \qquad (27\text{-}2)$$

For an *RL* low-pass filter, the cutoff frequencies are

$$\omega_C = \frac{1}{\tau} = \frac{R}{L} \qquad (27\text{-}3)$$

and

$$f_C = \frac{\omega_C}{2\pi} = \frac{R}{2\pi L} \qquad (27\text{-}4)$$

The frequency response of a filter circuit is generally shown on two semi-logarithmic graphs where the abscissa (horizontal axis) for each graph gives the frequency on a logarithmic scale. The ordinate (vertical axis) of one graph shows the voltage gain in decibels, while the ordinate of the other graph shows the phase shift of the output voltage with respect to the applied input voltage. The voltage gain and phase shift for any frequency are determined by finding the *transfer function TF* of the given filter. The transfer function is defined as the ratio of the output voltage phasor to the input voltage phasor.

$$\mathbf{TF} = \frac{\mathbf{V}_{out}}{\mathbf{V}_{in}} \qquad (27\text{-}5)$$

CALCULATIONS

The *RC* Low-Pass Filter

1. Write the transfer function $\mathbf{TF} = \mathbf{V}_{out}/\mathbf{E}$ for the *RC* low-pass filter of Figure 27-2.

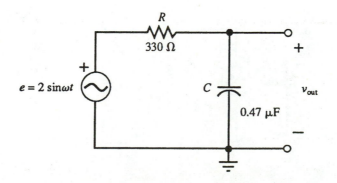

Figure 27-2 RC low-pass filter

2. Determine the cutoff frequencies (in radians per second and in hertz) of the circuit in Figure 27-2. Record the values here.

ω_C		rad/s
f_C		Hz

3. Sketch the straight-line approximations of the frequency responses (A_v in dB versus frequency and θ versus frequency) for the *RC* low-pass filter on Graph 27-1.

The *RL* Low-Pass Filter

4. Write the transfer function **TF = V_{out}/E** for the *RL* low-pass filter of Figure 27-3.

5. Determine the cutoff frequencies (in radians per second and in hertz) of the circuit in Figure 27-3. Record the values here.

ω_C		rad/s
f_C		Hz

6. Sketch the straight-line approximations of the frequency responses (A_v in dB versus frequency and θ versus frequency) for the *RL* low-pass filter on Graph 27-2.

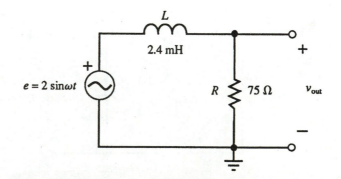

Figure 27-3 RL low-pass filter

MEASUREMENTS

The *RC* Low-Pass Filter

7. Assemble the circuit shown in Figure 27-2. Connect Ch1 of the oscilloscope to the output of the signal generator and adjust the output to have an amplitude of 2.0 V_p (4.0 $V_{p\text{-}p}$) at a frequency of $f = 100$ Hz.

8. Use Ch1 as the trigger source and connect Ch2 of the oscilloscope across the capacitor C. Measure the amplitude and phase angle (with respect to e) of the sinusoidal output voltage v_{out}. Enter the results in Table 27-2.

9. Increase the frequency of the signal generator to the frequencies indicated in Table 27-2. If necessary, adjust the amplitude of the signal generator to ensure that the output is maintained at 2.0 V_p (4.0 $V_{p\text{-}p}$). Measure the amplitude and phase shift of v_{out} (with respect to e) for each frequency.

f	v_{out}	
	Amplitude	Phase shift
100 Hz		
200 Hz		
400 Hz		
800 Hz		
1 kHz		
2 kHz		
4 kHz		
8 kHz		
10 kHz		

Table 27-2

10. Determine the cutoff frequency by adjusting the frequency of the generator until the output voltage has an amplitude of $V_{out} = (0.707)(2.0\ V_p) = 1.41\ V_p$. Record the measured cutoff frequency below.

f_C	

The *RL* Low-Pass Filter

11. Construct the circuit shown in Figure 27-3. Connect Ch1 of the oscilloscope to the output of the signal generator and adjust the output to have an amplitude of 2.0 V_p (4.0 V_{p-p}) at a frequency of $f = 100$ Hz.

12. Use Ch1 as the trigger source and connect Ch2 of the oscilloscope across the resistor R. Measure the amplitude and phase angle (with respect to e) of the sinusoidal output voltage v_{out}. Enter the results in Table 27-3.

13. Increase the frequency of the signal generator to the frequencies indicated in Table 27-3. If necessary, adjust the amplitude of the signal generator to ensure that the output is maintained at 2.0 V_p (4.0 V_{p-p}). Measure the amplitude and phase shift of v_{out} (with respect to e) for each frequency.

f	v_{out}	
	Amplitude	Phase shift
100 Hz		
200 Hz		
400 Hz		
800 Hz		
1 kHz		
2 kHz		
4 kHz		
8 kHz		
10 kHz		

Table 27-3

14. Determine the cutoff frequency by adjusting the frequency of the generator until the output voltage has an amplitude of $V_{out} = (0.707)(2.0 \, V_p) = 1.41 \, V_p$. Record the measured cutoff frequency here.

f_C	

CONCLUSIONS

The *RC* Low-Pass Filter

15. Use your measurements recorded in Table 27-2 to calculate the magnitude of the gain as a ratio of the amplitudes V_{out}/V_{in}. Calculate the voltage gain in decibels as

$$[A_v]_{dB} = 20 \log \frac{V_{out}}{V_{in}} \qquad (27\text{-}6)$$

Record the calculated voltage gain and measured phase shift for each frequency in Table 27-4.

f	$A_v = V_{out}/V_{in}$	$[A_v]_{dB}$	θ
100 Hz			
200 Hz			
400 Hz			
800 Hz			
1 kHz			
2 kHz			
4 kHz			
8 kHz			
10 kHz			

Table 27-4

16. Plot the data of Table 27-4 on the semi-logarithmic scales of Graph 27-1 (page 217). Connect the points with the best smooth continuous curve. You should find that the actual response is closely predicted by the straight-line approximations of the Bode plot.

17. Compare the measured cutoff frequency f_C of Step 10 to the theoretical value predicted by the transfer function in Step 2.

The *RL* Low-Pass Filter

18. Use your measurements recorded in Table 27-3 to calculate the magnitude of the gain as a ratio of the amplitudes V_{out}/V_{in}. Determine the voltage gain in decibels. Record the calculated voltage gain and measured phase shift for each frequency in Table 27-5.

f	$A_v = V_{out}/V_{in}$	$[A_v]_{dB}$	θ
100 Hz			
200 Hz			
400 Hz			
800 Hz			
1 kHz			
2 kHz			
4 kHz			
8 kHz			
10 kHz			

Table 27-5

19. Plot the data of Table 27-5 on the semi-logarithmic scales of Graph 27-2 (page 218). Connect the points with the best smooth continuous curve. You should find that the actual response is closely predicted by the straight-line approximations of the Bode plot.

20. Compare the measured cutoff frequency f_C of Step 14 to the theoretical value predicted by the transfer function in Step 5.

_____

_____

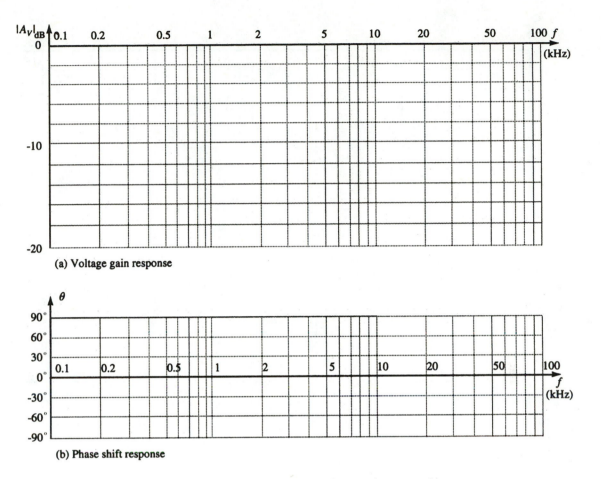

(a) Voltage gain response

(b) Phase shift response

Graph 27-1 Frequency response of an RC low-pass filter

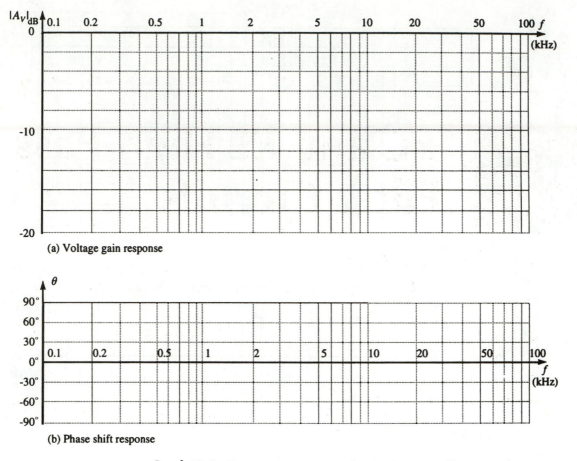

(a) Voltage gain response

(b) Phase shift response

Graph 27-2   Frequency response of an RL low-pass filter

LAB 28

RC and *RL* High-Pass Filter Circuits

OBJECTIVES

After completing this lab, you will be able to
- develop the transfer function for a high-pass filter circuit,
- determine the cutoff frequency of a high-pass filter circuit,
- sketch the Bode plot of the transfer fuction for a high-pass filter,
- compare the measured voltage gain response of a high-pass filter to the theoretical asymptotic response predicted by a Bode plot,
- explain why the voltage gain of a high-pass filter increases at a rate of 20 dB/decade below the cutoff frequency.

EQUIPMENT REQUIRED

☐ Dual trace oscilloscope
☐ Signal generator (sinusoidal function generator)
☐ DMM
Note: Record this equipment in Table 28-1.

COMPONENTS

☐ Resistors: 75-Ω, 330-Ω (1/4-W carbon, 5% tolerance)
☐ Capacitors: 0.47-μF (10% tolerance)
☐ Inductors: 2.4-mH (iron core, 5% tolerance)

EQUIPMENT USED

Instrument	Manufacturer/Model No.	Serial No.
Oscilloscope		
Signal generator		

Table 28-1

TEXT REFERENCE

Section 22.5   THE HIGH-PASS FILTER CIRCUIT

DISCUSSION

As the name implies, the *high-pass filter* permits high frequencies to pass from the input through to the output of the filter. Due to the abundance of electric motors and fluorescent lights, *60-Hz noise* is the most prevalent unwanted signal around us. Although many applications exist, one of the most common uses for the high-pass filter is to prevent 60-Hz noise from entering a sensitive electrical or electronic system. Figure 28-1 shows both *RC* and *RL* high-pass filters.

As in the low-pass filter circuit, the *cutoff frequency* f_C is that frequency at which the amplitude of the output voltage is 0.707 of the maximum amplitude. However, in the case of high-pass filters, the maximum amplitude occurs for high frequencies rather than for low frequencies. Since this frequency corresponds to half power, the cutoff frequency is also called the *half-power* or *3-dB down frequency*.

The cutoff frequencies for an *RC* high-pass filter are identical to the values for the low-pass *RC* filter and are given as follows:

$$\omega_C = \frac{1}{\tau} = \frac{1}{RC} \tag{28-1}$$

and

$$f_C = \frac{\omega_C}{2\pi} = \frac{1}{2\pi RC} \tag{28-2}$$

Similarly, for an *RL* high-pass filter, the cutoff frequencies are

$$\omega_C = \frac{1}{\tau} = \frac{R}{L} \tag{28-3}$$

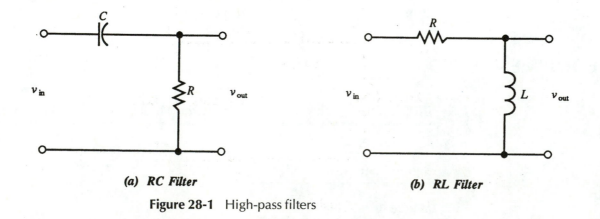

(a) RC Filter   *(b) RL Filter*

Figure 28-1   High-pass filters

and

$$f_C = \frac{\omega_C}{2\pi} = \frac{R}{2\pi L} \qquad (28\text{-}4)$$

As in the low-pass filter, the frequency response of a high-pass filter circuit is generally shown on two semi-logarithmic graphs where the abscissa (horizontal axis) for each graph gives the frequency on a logarithmic scale. The ordinate (vertical axis) of one graph shows the voltage gain in decibels, while the ordinate of the other graph shows the phase shift of the output voltage with respect to the applied input voltage. The voltage gain and phase shift for any frequency are determined by finding the *transfer function TF* of the given filter. Recall that the transfer function is defined as the ratio of the output voltage phasor to the input voltage phasor.

$$\mathbf{TF} = \frac{\mathbf{V}_{out}}{\mathbf{V}_{in}} \qquad (28\text{-}5)$$

CALCULATIONS

The *RC* High-Pass Filter

1. Write the transfer function **TF** = $\mathbf{V}_{out}/\mathbf{E}$ for the *RC* high-pass filter of Figure 28-2.

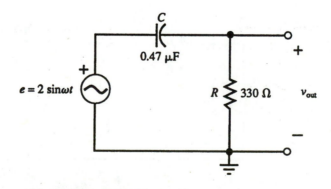

Figure 28-2 *RC* high-pass filter

2. Determine the cutoff frequencies (in radians per second and in hertz) of the circuit in Figure 28-2. Record the values here.

ω_C		rad/s
f_C		Hz

3. Sketch the straight-line approximations of the frequency responses (A_v in dB versus frequency and θ versus frequency) for the *RC* high-pass filter on Graph 28-1 (page 226).

The *RL* High-Pass Filter

4. Write the transfer function **TF** = $\mathbf{V}_{out}/\mathbf{E}$ for the *RL* high-pass filter of Figure 28-3.

5. Determine the cutoff frequencies (in radians per second and in hertz) of the circuit in Figure 28-3. Record the values here.

ω_C		rad/s
f_C		Hz

6. Sketch the straight-line approximations of the frequency responses (A_v in dB versus frequency and θ versus frequency) for the *RL* high-pass filter on Graph 28-2 (page 227).

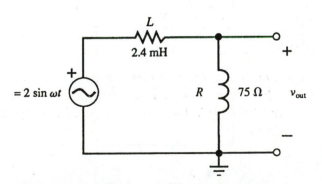

Figure 28-3 *RL* high-pass filter

MEASUREMENTS

The *RC* High-Pass Filter

7. Assemble the circuit shown in Figure 28-2. Connect Ch1 of the oscilloscope to the output of the signal generator and adjust the output to have an amplitude of 2.0 V_p (4.0 V_{p-p}) at a frequency of $f = 100$ Hz.

8. Use Ch1 as the trigger source and connect Ch2 of the oscilloscope across the resistor R. Measure the amplitude and phase angle (with respect to e) of the sinusoidal output voltage v_{out}. Enter the results in Table 28-2.

9. Increase the frequency of the signal generator to the frequencies indicated in Table 28-2. If necessary, adjust the amplitude of the signal generator to ensure that the output is maintained at 2.0 V_p (4.0 V_{p-p}). Measure the amplitude and phase shift of v_{out} (with respect to e) for each frequency.

f	v_{out}	
	Amplitude	Phase shift
100 Hz		
200 Hz		
400 Hz		
800 Hz		
1 kHz		
2 kHz		
4 kHz		
8 kHz		
10 kHz		

Table 28-2

10. Determine the cutoff frequency by adjusting the frequency of the generator until the output voltage has an amplitude of $V_{out} = (0.707)(2.0 \ V_p) = 1.41 \ V_p$. Record the measured cutoff frequency below.

f_C	

The *RL* High-Pass Filter

11. Construct the circuit shown in Figure 28-3. Connect Ch1 of the oscilloscope to the output of the signal generator and adjust the output to have an amplitude of 2.0 V_p (4.0 V_{p-p}) at a frequency of $f = 100$ Hz.

12. Use Ch1 as the trigger source and connect Ch2 of the oscilloscope across the inductor L. Measure the amplitude and phase angle (with respect to e) of the sinusoidal output voltage v_{out}. Enter the results in Table 28-3.

13. Increase the frequency of the signal generator to the frequencies indicated in Table 28-3. If necessary, adjust the amplitude of the signal generator to ensure that the output is maintained at 2.0 V_p (4.0 V_{p-p}). Measure the amplitude and phase shift of v_{out} (with respect to e) for each frequency.

f	v_{out}	
	Amplitude	Phase shift
100 Hz		
200 Hz		
400 Hz		
800 Hz		
1 kHz		
2 kHz		
4 kHz		
8 kHz		
10 kHz		

Table 28-3

14. Determine the cutoff frequency by adjusting the frequency of the generator until the output voltage has an amplitude of $V_{out} = (0.707)(2.0\ V_p) = 1.41\ V_p$. Record the measured cutoff frequency below.

f_C	

CONCLUSIONS

The *RC* High-Pass Filter

15. Use the measurements recorded in Table 28-2 to calculate the magnitude of the gain as a ratio of the amplitudes V_{out}/V_{in}. Calculate the voltage gain in decibels as

$$[A_v]_{dB} = 20 \log \frac{V_{out}}{V_{in}} \qquad (28\text{-}6)$$

Record the calculated voltage gain and measured phase shift for each frequency in Table 28-4.

f	$A_v = V_{out}/V_{in}$	$[A_v]_{dB}$	θ
100 Hz			
200 Hz			
400 Hz			
800 Hz			
1 kHz			
2 kHz			
4 kHz			
8 kHz			
10 kHz			

Table 28-4

16. Plot the data of Table 28-4 on the semi-logarithmic scales of Graph 28-1 (next page). Connect the points with the best smooth continuous curve. You should find that the actual response is closely approximated by the Bode plot.

17. Compare the measured cutoff frequency f_C of Step 10 to the theoretical value predicted by the transfer function in Step 2.

_____

_____

The *RL* High-Pass Filter

18. Use the measurements recorded in Table 28-3 to calculate the magnitude of the gain as a ratio of the amplitudes V_{out}/V_{in}. Determine the voltage gain in decibels. Record the calculated voltage gain and measured phase shift for each frequency in Table 28-5.

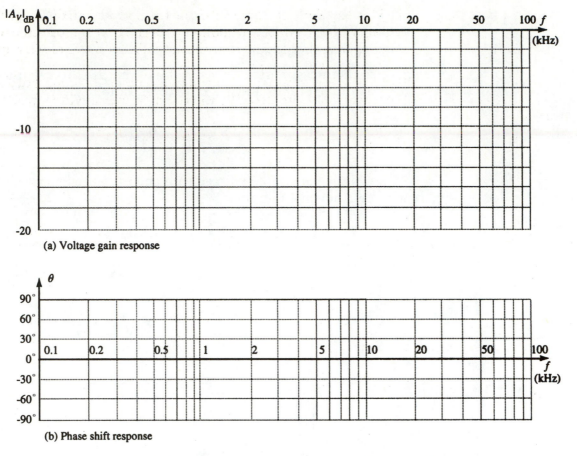

(a) Voltage gain response

(b) Phase shift response

Graph 28-1 Frequency response of an RC high-pass filter

f	$A_v = V_{out}/V_{in}$	$[A_v]_{dB}$	θ
100 Hz			
200 Hz			
400 Hz			
800 Hz			
1 kHz			
2 kHz			
4 kHz			
8 kHz			
10 kHz			

Table 28-5

19. Plot the data of Table 28-5 on the semi-logarithmic scales of Graph 28-2. Connect the points with the best smooth continuous curve. You should find that the actual response is closely approximated by the Bode plot.

20. Compare the measured cutoff frequency f_C of Step 14 to the theoretical value predicted by the transfer function in Step 5.

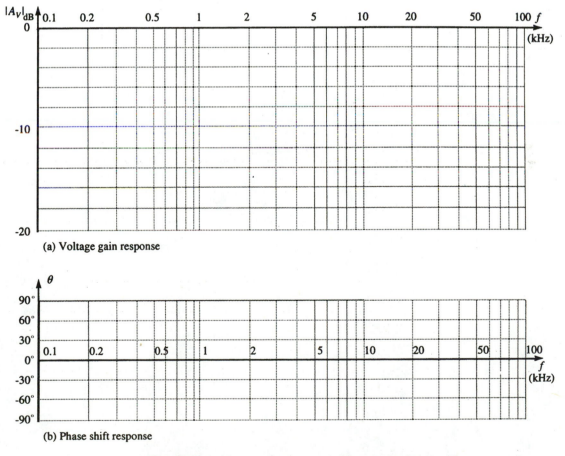

(a) Voltage gain response

(b) Phase shift response

Graph 28-2 Frequency response of an *RL* high-pass filter

Bandpass Filter

OBJECTIVES

After completing this lab, you will be able to
- calculate the cutoff frequencies of a bandpass filter by examining the individual low-pass and high-pass stages of a filter,
- derive the transfer function for each stage of a bandpass filter,
- sketch the Bode plot of a bandpass filter from the transfer function of the individual stages,
- explain why the slopes of the voltage gain response is 20 dB/decade on each side of the cutoff frequencies.

EQUIPMENT REQUIRED

☐ Dual trace oscilloscope
☐ Signal generator (sinusoidal function generator)
☐ DMM
Note: Record this equipment in Table 29-1.

COMPONENTS

☐ Resistors: 330-Ω (2) (1/4-W carbon, 5% tolerance)
☐ Capacitors: 0.047-μF, 0.47-μF (10% tolerance)

EQUIPMENT USED

Instrument	Manufacturer/Model No.	Serial No.
Oscilloscope		
Signal generator		

Table 29-1

TEXT REFERENCE

Section 22.6   BANDPASS FILTER

DISCUSSION

Bandpass filters permit a range of frequencies to pass from one stage to another. Many different types of bandpass filters are used throughout electronics. For instance, the typical television receiver uses several stages of filtering to achieve a bandwidth of 6 MHz, while an AM receiver has circuitry which restricts the bandwidth to only 10 kHz. Although bandpass filters can be very elaborate, depending on the application, they can also be constructed simply by combining a low-pass filter and a high-pass filter as shown in Figure 29-1. Since expense and other design difficulties make *RL* filters impractical, *RC* filters are used almost exclusively.

The bandpass filter of Figure 29-1 has two cutoff frequencies as determined by the cutoff frequencies of the individual stages. The frequency below which the high-pass filter attenuates the signal is called the *lower cutoff frequency* f_1. The frequency at which the low-pass filter begins to attenuate is called the *upper cutoff frequency* f_2. As expected, the difference between the two frequencies is the *bandwidth BW*. In order for the circuit to operate predictably, the cutoff frequencies should be separated by at least one *decade* and the first stage must be the high-pass circuit.

The lower cutoff frequency ω_1 in radians per second is found as

$$\omega_1 = \frac{1}{\tau_1} = \frac{1}{R_1 C_1} \tag{29-1}$$

with a corresponding frequency f_1 in hertz,

$$f_1 = \frac{\omega_1}{2\pi} = \frac{1}{2\pi R_1 C_1} \tag{29-2}$$

The upper cutoff frequency ω_2 in radians per second is

$$\omega_2 = \frac{1}{\tau_2} = \frac{1}{R_2 C_2} \tag{29-3}$$

with a corresponding frequency f_2 in hertz,

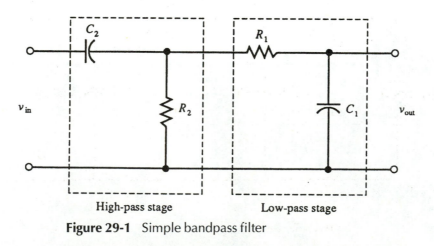

Figure 29-1   Simple bandpass filter

$$f_2 = \frac{\omega_2}{2\pi} = \frac{1}{2\pi R_2 C_2} \qquad (29\text{--}4)$$

The bandwidth of the filter is determined as

$$BW = f_2 - f_1 \qquad (29\text{-}5)$$

CALCULATIONS

1. Calculate and record the cutoff frequency (in radians per second and in hertz) for the high-pass stage in the circuit of Figure 29-2.

ω_1		rad/s
f_1		Hz

2. Calculate and record the cutoff frequency (in radians per second and in hertz) for the low-pass stage in the circuit of Figure 29-2.

ω_2		rad/s
f_2		Hz

3. Calculate and record the bandwidth of the filter circuit of Figure 29-2.

BW		Hz

4. From the calculations of Steps 1 and 2 sketch the Bode plots for the bandpass filter of Figure 29-2. Use the semilogarithmic scales of Graph 29-1 (page 234).

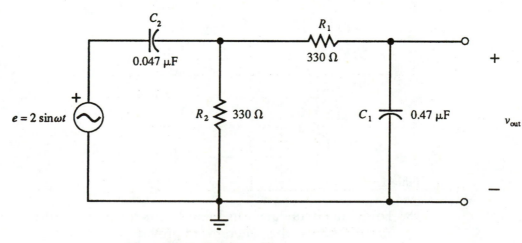

Figure 29-2 Bandpass filter

MEASUREMENTS

5. Assemble the circuit shown in Figure 29-2. Connect Ch1 of the oscilloscope to the output of the signal generator and adjust the output to have an amplitude of 2.0 V_p (4.0 V_{p-p}) at a frequency of f = 100 Hz.
6. Use Ch1 as the trigger source and connect Ch2 of the oscilloscope across the output of the circuit. Measure the amplitude and phase angle of the output voltage v_{out} (with respect to e). Enter the results in Table 29-2.
7. Increase the frequency of the signal generator to the frequencies indicated in Table 29-2. Due to loading effects of the circuit, it will be necessary to readjust the amplitude of the signal generator at each frequency to ensure that the output is maintained at 2.0 V_p (4.0 V_{p-p}). Measure and record the amplitude and phase shift of the output voltage v_{out} (with respect to e) for each frequency.

f	v_{out}	
	Amplitude	Phase shift
100 Hz		
200 Hz		
400 Hz		
800 Hz		
1 kHz		
2 kHz		
4 kHz		
8 kHz		
10 kHz		
20 kHz		
40 kHz		
80 kHz		
100 kHz		

Table 29-2

8. For which frequency f in Table 29-2 is the output voltage a maximum? Adjust the generator to provide an output of 2.0 V_p (4.0 V_{p-p}) at this frequency. While observing the oscilloscope, adjust

the the generator frequency until voltage V_{out} is at the maximum value. Measure and record the center frequency f_O and the corresponding amplitude V_{out}.

f_o	

9. Decrease the frequency until the output voltage V_{out} is reduced to 0.707 of the maximum value found in Step 8. Record the lower half-power frequency f_1 and the corresponding output voltage V_{out}. (Ensure that the output of the signal generator is at 2.0 V_p.)

f_1	

10. Increase the frequency above the center frequency until the output voltage V_{out} is again reduced to 0.707 of the maximum value found in Step 8. Record the upper half-power frequency f_2 and the corresponding resistor voltage V_{out}. (Ensure that the output of the signal generator is at 2.0 V_p.)

f_2	

CONCLUSIONS

11. For each voltage measurement in Table 29-2, determine the voltage gain as both a ratio of amplitudes $A_v = V_{out}/E$ and in decibels as

$$[A_v]_{dB} = 20 \log \frac{V_{out}}{E} \tag{29-6}$$

    Record the calculated voltage gain and measured phase shift for each frequency in Table 29-3 (next page).

12. Plot the data of Table 29-3 on the semi-logarithmic scale of Graph 29-1. Connect the points with the best smooth continuous curve.

13. Compare the measured cutoff frequencies of Step 9 and 10 to the theoretical frequencies determined in Steps 1 and 2.

_____

_____

14. How do the actual voltage gain and phase shift responses of the output voltage compare to the predicted responses?

_____

_____

_____

f	$A_v = V_{\text{out}}/E$	$[A_v]_{\text{dB}}$	θ
100 Hz			
200 Hz			
400 Hz			
800 Hz			
1 kHz			
2 kHz			
4 kHz			
8 kHz			
10 kHz			
20 kHz			
40 kHz			
80 kHz			
100 kHz			

Table 29-3

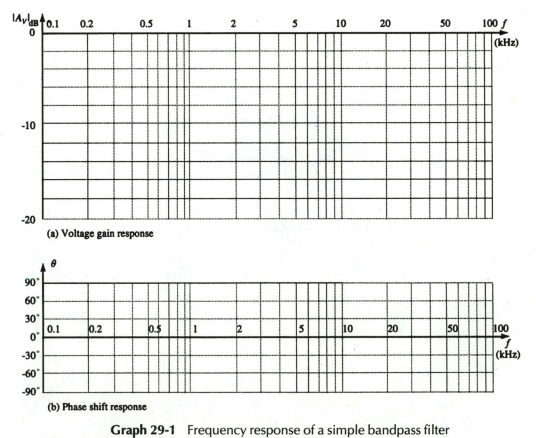

(a) Voltage gain response

(b) Phase shift response

Graph 29-1   Frequency response of a simple bandpass filter

Voltages and Currents in Balanced Three-Phase Systems

OBJECTIVES

After completing this lab, you will be able to
- draw the three-phase time varying voltage waveforms,
- verify line and phase voltage relationships for a Y-load,
- verify line and phase current relationships for a Δ-load,
- verify the single-phase equivalent method of analysis.

EQUIPMENT REQUIRED

☐ Dual channel oscilloscope
☐ ac ammeter (A DMM with a 2-A range is adequate)
☐ DMM
☐ Phase angle meter (optional)

POWER SUPPLY

☐ Three-phase 120/208 V (60 Hz), preferably variable such as by means of a 3-phase autotransformer

COMPONENTS

☐ Resistors: 100-Ω, 200-W (three required), 250-Ω, 200-W (three required)

EQUIPMENT USED

Instrument	Manufacturer/Model No.	Serial No.
Oscilloscope		
ac Ammeter		
DMM		

Table 30-1

TEXT REFERENCE

Section 23.1   THREE-PHASE VOLTAGE GENERATION
Section 23.3   BASIC THREE-PHASE RELATIONSHIPS

DISCUSSION

A balanced three-phase system is a system where source voltages are equal in magnitude and phase-displaced from each other by 120°, and the impedance in each phase is identical. Most electrical utilities are three-phase.

Y-Loads. For a Y-Load, Figure 30-1, line-to-line voltages are $\sqrt{3}$ times line-to-neutral voltages and each line-to-line voltage leads its corresponding line-to-neutral voltage by 30°. For example,

$$\mathbf{V}_{ab} = \sqrt{3}\ \mathbf{V}_{an}\angle 30° \tag{30-1}$$

Current in the neutral of a balanced system (if a neutral line is present) is zero.

Δ-Loads. The magnitude of line current for a balanced Δ-Load, Figure

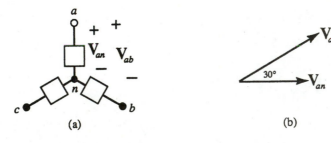

(a)   (b)

(a) Circuit   *(b) Voltage relationships*

Figure 30-1   A Y-load

30-2, is $\sqrt{3}$ times the magnitude of the phase current and each line current lags its corresponding phase current by 30°. Thus,

$$\mathbf{I}_a = \sqrt{3}\ \mathbf{I}_{ab}\angle -30° \qquad (30\text{-}2)$$

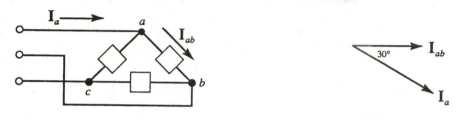

(a) Circuit

(b) Current relationships

Figure 30-2 A Δ-load

Equivalent Y and Δ-Loads. For purposes of analysis, a balanced Δ-load may be replaced by an equivalent balanced Y-load with

$$\mathbf{Z}_Y = \mathbf{Z}_\Delta/3 \qquad (30\text{-}3)$$

Single-Phase Equivalent. Since all three phases of a balanced system are identical, any one phase can be taken to represent the behavior of all. This permits you to reduce a circuit to its *single-phase equivalent* (which you can call *Phase a*), solve for voltages and currents for that phase, then determine those for the other two phases by inspection—i.e., by shifting all angles by –120° to get *Phase b* quantities and by 120° to get *Phase c* quantities. This is the method used in practice.

Safety Note

In this lab, you will be working with 120/208 VAC. Such voltages are dangerous and you must be aware of and observe safety precautions. Familiarize yourself with your laboratory's safety practices. **Always ensure that power is off when you are assembling, changing or otherwise working on your circuit.** If possible, use a variable ac power supply (such as a three-phase variable autotransformer) as the source. Start at zero volts and gradually increase the voltage while watching the meters for signs of trouble. In any event, have your instructor check your circuit before you energize it. **USE EXTREME CAUTION IN THIS LAB.**

MEASUREMENTS

PART A: Three-Phase Sinusoidal Voltages

1. With power off, connect the scope grounds to the ground point of the

> **General Safety Notes**
>
> 1. With power off, assemble your circuit and have your lab partner double check it.
>
> *(continues on next page)*

source, Figure 30-3. Use ×10 probes and set the vertical sensitivity high enough so as not to damage the oscilloscope.

a. Connect Ch1 to view voltage e_{AN} and Ch2 to view e_{BN}. (Trigger on Ch1, positive slope.) Set both channels for the same sensitivity. Energize the circuit. (For this test, you need only a relatively low voltage (say 50 V) to get adequate waveforms.) Sketch voltages in Figure 30-4(a) (next page). Use voltage e_{AN} as reference.

b. De-energize the circuit, and with Ch1 still connected, move Ch2's probe to view voltage e_{CN} and sketch on Figure 30-4(b), being careful to note the phase relationship.

c. Draw the phasor diagram for the above waveforms.

2. Have your instructor check the circuit before you energize it. If you are using a Powerstat or other variable ac source, gradually increase voltage from zero, watching the meters for signs of trouble.

3. Turn power off before changing your circuit for the next test.

Check with your instructor for specific safety instructions.

PART B: Line and Neutral Voltages for a Y

2. Remove the oscilloscope from Figure 30-3.

a. Using a DMM or VOM, set E_{AN} to 120 V, then measure the magnitude of each line-to-neutral voltage and record in Table 30-2.

b. Repeat for each line-to-line voltage.

c. If you have a phase angle meter, measure the phase shift between $\mathbf{E}_{AN}$ and $\mathbf{E}_{BN}$ and between $\mathbf{E}_{AN}$ and $\mathbf{E}_{AB}$. Ask your instructor for connection details. Record here.

E_{AN}	
E_{BN}	
E_{CN}	
E_{AB}	
E_{BC}	
E_{CA}	

Table 30-2

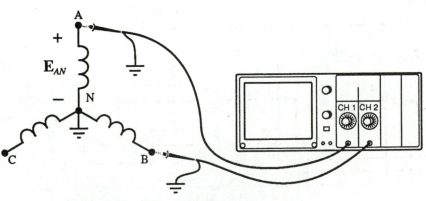

Figure 30-3   Circuit for Test 1

(a) Waveforms   *(b) Phasors*

Figure 30-4   Waveforms and phasor diagram for Test 1

d. What is the theoretical relationship between E_{AB} and E_{AN}? Do the measurements confirm this?

PART C: Currents for a Y-Load

3. Connect the circuit of Figure 30-5. If you have enough ammeters, use one for each line. Otherwise, measure current in *line a*, turn power off, move the meter to *line b*, measure current and so on. (The ammeter may be a standard ac meter or the ac current range of a DMM. For the load resistance shown, a 2-A range is adequate.)
   a. Measure each line current and record in Table 30-3.
   b. Measure the neutral current and record.

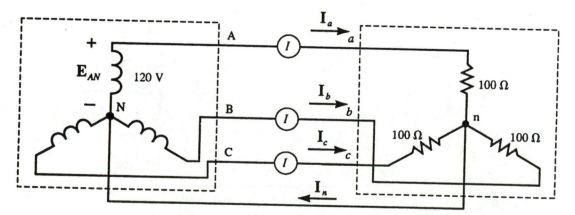

Figure 30-5   Circuit for Test 3. Each resistor is rated 200-W.

c. Turn power off and remove the neutral conductor from between n and N. Measure the line currents again. Are they the same as in Test 2(a)?

I_a	
I_b	
I_c	
I_n	

d. Using the measured source voltage and load resistance, compute the magnitude of the line current and compare to the measured value.

Table 30-3

PART D: Currents for a Δ-Load

4. Assemble the circuit of Figure 30-6, using 250-Ω resistors. Measure line current I_a and phase current I_{ab}.

$I_a = $ _____ $I_{ab} = $ _____

What is the theoretical relationship between I_a and I_{ab}? Do the measurements confirm this?

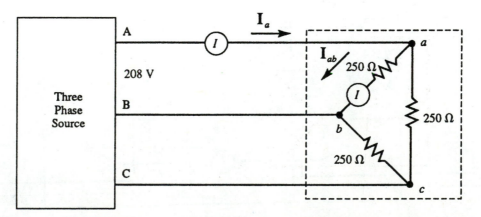

Figure 30-6   Circuit for Test 4. For three-phase source, see Figure 30-5.

PART E: The Single Phase Equivalent

5. Consider the three-phase circuit of Figure 30-7(a). (Do not assemble this circuit.) Each resistor of the Y is 250 Ω and each resistor of the Δ is 300 Ω. Source voltage is 120-V, line-to-neutral.
   a. Convert the Δ load to a Y-equivalent and convert the circuit to its single-phase equivalent. Sketch as Figure 30-7(b).
   b. Assemble the single-phase equivalent. Set the source voltage E_{AN} to 120 V and measure line current.

   E_{AN} = _____ I_A = _____

   c. Analyze the single-phase equivalent and compare calculated current I_A to that measured in (b).

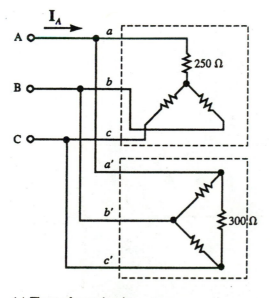

(a) Three-phase circuit   *(b) Single phase equivalent*

Figure 30-7 Circuit for Test 5

PROBLEMS

6. Assume each resistor of the Δ-load of Figure 30-7(a) is replaced with a 10-μF capacitor.
   a. Convert the circuit to its single-phase equivalent. Sketch in the space below.

   b. Calculate the magnitude of the line current I_A.
   c. Calculate the magnitude of the phase current in the Δ-load.

Power in Three-Phase Systems

OBJECTIVE

After completing this lab, you will be able to
* Measure power in a three-phase system.

EQUIPMENT REQUIRED

☐ Single-phase wattmeters (2 required)
☐ ac ammeter (A DMM with a 2-A range is adequate)
☐ DMM

POWER SUPPLY

☐ Three-phase 120/208 V (60 Hz), preferably variable such as by means of a 3-phase autotransformer

COMPONENTS

☐ Resistors: 100-Ω, 200-W (3 required), 250-Ω, 200-W (3 required)
☐ Capacitors: 10-μF (non-electrolytic), rated for operation at 120-VAC (3 required)

Safety Note. In this lab, you will be working with 120/208 VAC. Such voltages are dangerous and you must be aware of and observe safety precautions. Familiarize yourself with your lab's safety practices. Always ensure that power is off when you are assembling, changing or otherwise working on your circuit. **USE EXTREME CAUTION AT ALL TIMES.**

EQUIPMENT USED

Instrument	Manufacturer/Model No.	Serial No.
Single-phase wattmeters		
ac Ammeter		
DMM		

Table 31-1

TEXT REFERENCE

Section 23.5   POWER IN A BALANCED SYSTEM
Section 23.6   MEASURING POWER IN THREE-PHASE CIRCUITS
Section 23.7   UNBALANCED SYSTEMS

DISCUSSION

Power in Three-Phase Systems. Power in a balanced three-phase system is three times the power to one phase. Power to one phase P_ϕ, is given by

$$P_\phi = V_\phi I_\phi \cos \theta_\phi \qquad (31\text{-}1)$$

where V_ϕ and I_ϕ are the magnitudes of the phase voltage and current and θ_ϕ, the angle between them, is the angle of the phase load impedance. This formula applies to both Y and Δ loads. Total power is three times this.

An alternate power formula based on line voltages and currents is

$$P_T = \sqrt{3} \; V_L I_L \cos \theta_\phi \qquad (31\text{-}2)$$

where V_L and I_L are the magnitudes of the line-to-line voltages and line currents respectively. This formula applies to both Y and Δ loads. Note that the angle in this formula is the angle of the load impedance—it is not the angle between $\mathbf{V}_L$ and $\mathbf{I}_L$.

For unbalanced systems, Equation 31-2 does not apply; power must be calculated on a per phase basis and added.

Measuring Power in Three-Phase Systems. Power to a three-phase load may be measured in several ways. If the system is balanced, you can measure power to one phase, then multiply by three. However, a more widely used method is to use (N–1) wattmeters where N is the number of wires in the system. Thus, for a 3-wire load, you need only two wattmeters, while for a 4-wire load, you need three wattmeters. Both connections (shown in Figure 31-1) measure total power for balanced or unbalanced loads.

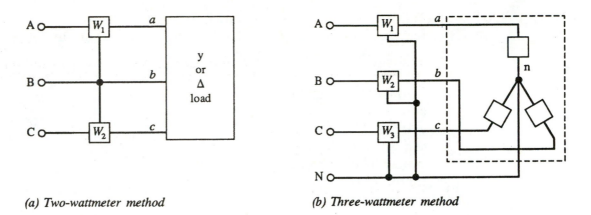

(a) Two-wattmeter method   (b) Three-wattmeter method

Figure 31-1   Measuring power in three-phase systems

For the two wattmeter method, one wattmeter reads higher than the other. If the load is balanced and its power factor angle exceeds 60°, the low reading wattmeter will read negative. In this case, subtract it from the high reading meter to get total power. The power factor for balanced loads may be determined from the watts-ratio curve of Figure 31-2 or by means of the formula

$$\tan \theta_{\phi} = \sqrt{3}\left(\frac{P_h - P_{\ell}}{P_h + P_{\ell}}\right)$$

(31-3)

where P_h and P_{ℓ} represent the high-reading and low-reading wattmeter values respectively. Once θ_{ϕ} is determined from Equation 31-3, the power factor can be determined from $F_p = \cos \theta_{\phi}$.

Wattmeter Connections. As noted in Lab 20, separate ± connections are provided on some wattmeters to permit alternate connections as

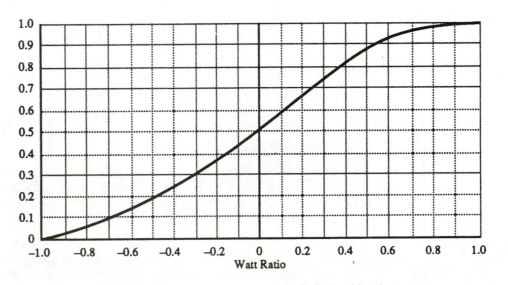

Figure 31-2   Watts-ratio curve for balanced loads

in Figure 20-1. If your meters are like this, connect as in Figure 20-1(a) for this lab. Otherwise, connect as appropriate to your meter. (If necessary ask your instructor for advice.)

What a Wattmeter Really Reads. A wattmeter reads the product of the magnitude of the voltage across its voltage circuit times the magnitude of the current through its current coil times the cosine of the angle between them. This angle is not the angle of the load impedance. For example, in Figure 31-3 (below), $W_1 = V_{ab} I_a \cos \theta_1$ where θ_1 is the angle between $\mathbf{V}_{ab}$ and $\mathbf{I}_a$, while $W_2 = V_{cb} I_c \cos \theta_2$ where θ_2 is the angle between $\mathbf{V}_{cb}$ and $\mathbf{I}_c$. Note that $\mathbf{V}_{cb} = \mathbf{V}_{bc}$.

General Safety Notes
1. With power off, assemble your circuit. Have your lab partner double check it.
2. Have your instructor check the circuit before you energize it. If you have a Powerstat or other variable ac source, gradually increase voltage from zero, watching the meters for signs of trouble.
3. Turn power off before changing your circuit for the next test.
Check with your instructor for specific safety instructions.

MEASUREMENTS

PART A: A Unity Power Factor Y-Load

1. a. Connect the circuit as in Figure 31-3. (The ammeter may be a standard ac ammeter or the ac current range on a DMM. For the component values used here, a 2-A range is adequate. Select a wattmeter to match—e.g., a wattmeter with a 300-W scale.)
    b. Measure voltage, current, and power and record in Table 31-2.

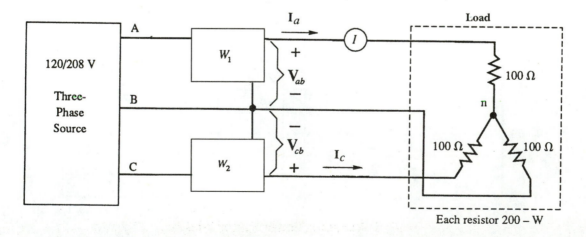

Figure 31-3   Circuit for Test 1

2. a. Analyze the circuit to determine P_T. Check by summing the wattmeter readings. How well do they agree?

V_{ab}	
I_a	
W_1	
W_2	

Table 31-2

b. Analyze the circuit and determine what wattmeter 1 should read. Compare it to the actual reading. Repeat for wattmeter 2.

PART B: A Leading Power Factor Y-Load

3. De-energize the circuit and add 10-µF of capacitance in parallel

with each resistor as in Figure 31-4. Measure voltage, current and power and record them in Table 31-3.

4. Power will remain unchanged (since the added capacitors dissipate no power). However, the power factor and hence the wattmeter readings change.

    a. Determine the angle θ of the load impedance, and from it, determine the load power factor.

    b. Determine the power factor using the watts-ratio curve.

        $F_p =$ _____.

    c. Analyze the circuit and determine the reading of each wattmeter. Compare to the measured values.

PART C: An Unbalanced Δ-Load

5. a. Assemble the load of Figure 31-5 (next page). Have the instruc-

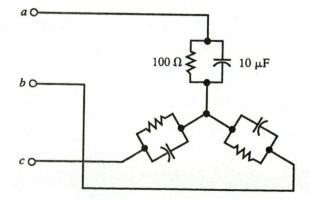

100 Ω   10 μF

Figure 31-4 Load for Test 2

V_{ab}	
I_a	
W_1	
W_2	

Table 31-3

tor check the circuit. Energize and measure voltage and power and record in Table 31-4.

   b. Determine total power by summing the wattmeter readings.
6. Determine P_T by analyzing the circuit and compare the results to those of Step 5(b). Use V_{ab} from Table 31-4 as the source voltage.

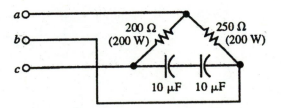

V_{ab}	
W_1	
W_2	

Table 31-4

Figure 31-5   Load for Test 3

PROBLEMS

7. Consider again the circuit of Figure 31-3. Change the load impedances to $Z_Y = 16\ \Omega\ \angle 65°$. Let $E_{AN} = 120\ V\ \angle 0°$.
   a. Determine total power to the load.
   b. Determine the reading of each wattmeter.
   c. Sum the wattmeter readings and show that they equal the result determined in (a).

LAB 32

The Iron-Core Transformer

OBJECTIVE

After completing this lab, you will be able to
- verify the turns ratio and phase relationships for a transformer,
- verify the concept of reflected impedance,
- determine the frequency response of an audio transformer,
- measure the regulation of a power transformer.

EQUIPMENT REQUIRED

☐ Oscilloscope
☐ DMM
☐ Signal or function generator

COMPONENTS

☐ Resistors: 10-Ω, 100-Ω (each 1/4 W), 10-Ω, 25-W
☐ Audio transformer (Hammond 145F or equivalent)
☐ 12.6-V filament transformer (Hammond 167L12 or equivalent)

EQUIPMENT USED

Instrument	Manufacturer/Model No.	Serial No.
Oscilloscope		
DMM		
Signal or function generator		

Table 32-1

TEXT REFERENCE

Section 24.2   THE IRON-CORE TRANSFORMER: THE IDEAL
                        MODEL
Section 24.3   REFLECTED IMPEDANCE
Section 24.8   VOLTAGE AND FREQUENCY EFFECTS

DISCUSSION

Iron-core transformers are used in power and audio frequency applications. To understand their behavior, we look first at the ideal transformer. An ideal transformer, Figure 32-1(a), is one that is characterized by its turns ratio

$$a = N_p/N_s \qquad (32\text{-}1)$$

where N_p and N_s are its primary and secondary turns respectively. Voltage and current ratios are given by

$$\mathbf{V}_p/\mathbf{V}_s = V_p/V_s = a \qquad (32\text{-}2)$$

$$\mathbf{I}_p/\mathbf{I}_s = I_p/I_s = 1/a \qquad (32\text{-}3)$$

Since an ideal transformer is lossless, apparent power out equals apparent power in. That is,

$$V_s I_s = V_p I_p \qquad (32\text{-}4)$$

Reflected Impedance. When a load impedance $\mathbf{Z}_L$ is connected to the secondary of an ideal transformer, Figure 32-1(c), it is seen in the primary as an impedance of

$$\mathbf{Z}_p = a^2\,\mathbf{Z}_L \qquad (32\text{-}5)$$

$\mathbf{Z}_p$ is the *reflected load impedance*. All impedances in the secondary (whether part of the load or not) are reflected in this fashion. This

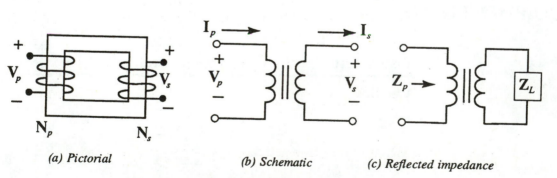

(a) Pictorial   *(b) Schematic*   *(c) Reflected impedance*

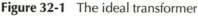

Figure 32-1   The ideal transformer

includes the secondary winding resistance of a real transformer, as you will see in Part C of this lab.

Phasing. Depending on the relative direction of windings, the secondary voltage is either in-phase or 180° out-of-phase with respect to the primary. Terminals that have the same polarity are referred to as corresponding terminals. Transformer terminals are often color coded to identify corresponding terminals; on schematic diagrams, corresponding terminals are marked with dots (or letters such as H_1, X_1).

Real Transformers. Real transformers differ from the above ideal in that they have primary and secondary winding resistances, leakage flux, core losses and a few other non-ideal characteristics. In many instances, these non-ideal characteristics can be neglected. In others, they cannot. In this lab, you will observe the effect of some of these. For example, you will find that the output voltage of a power transformer changes from no-load to full-load, resulting in an undesirable load voltage regulation. For an audio transformer, you will find that its output characteristics change over its bandwidth, resulting in a flat (ideal) response over its center range, but falling off at both its low frequency and high frequency ends. (This same effect occurs with power transformers, but since they are operated at one frequency only, it is unimportant.)

MEASUREMENTS

PART A: Turns Ratio

The turns ratio of a real transformer is equal to the ratio of its primary voltage to its secondary voltage at no-load. Since this is not the same as its voltage ratio under load (because of internal voltage drops), you must be careful when reading data sheets since some list voltage ratios under load, while others list no-load voltage ratios. For example, the Hammond 167L12 suggested for this lab has a no-load specification of 115/13.8 V but a full load specification of 115/12.6 V.

1. Consider Figure 32-2. Using the signal generator, apply a 60 Hz sine wave of about 6 V (rms) to the 115 V side of the 12.6 V transformer. (For safety reasons, we are applying a low voltage here, rather than the rated voltage of 115 V.)

   a. Measure primary and secondary voltages. $V_p =$ _____;

      $V_s =$ _____

      Determine the turns ratio using Equation 32-2.

      $a_{(measured)} =$ _____

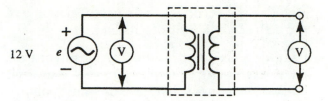

Figure 32-2 Circuit for Test 1. Use a low-voltage signal generator as the source.

b. Determine the turns ratio from its no-load rating and compare
    it to that of Test 1(a). $a_{\text{(Name-plate)}}$ = _____

PART B: Phase Relationships

2. Using the audio transformer, assemble the circuit of Figure 32-3.
    Set the source to a 1-kHz sine wave with an amplitude of 8 V (i.e.,
    16 $V_{\text{p-p}}$). Mark the transformer terminals for reference in deter-
    mining phase relationships.
    a. Measure the magnitude and phase of v_{cd} relative to the primary
        voltage and sketch in Figure 32-4 (next page). Now reverse the
        secondary scope leads and sketch v_{dc}. Based on these observa-
        tions, add the missing dot to the transformer of Figure 32-3. Ex-
        plain how you determined the position of this dot.

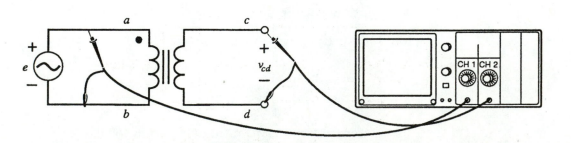

Figure 32-3 Circuit for Test 2

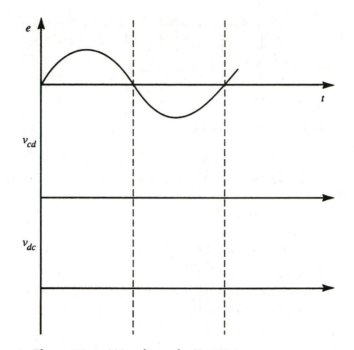

Figure 32-4 Waveforms for Test 2(a)

b. Based on the measurements of Test 2(a), what is the turns ratio

for this transformer? $a = $ _____

PART C: Reflected Impedance

3. Measure the 10-Ω sensing resistor and the 100-Ω load resistor, then assemble the circuit of Figure 32-5, using the audio transformer. Connect probes as shown.

$R_{sen} = $ _____ $R_L = $ _____

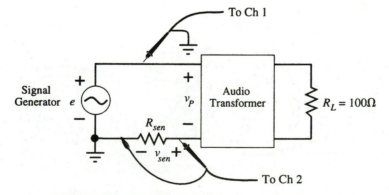

Figure 32-5 Circuit for Test 3

a. Using the differential mode of the oscilloscope, set the input voltage V_p of the transformer to 16 V peak-to-peak, 1-kHz.

b. Select Ch2 and measure the voltage across the sensing resistor. From this, calculate the primary current.

V_{sen} = _____ I_p = _____

c. Using the results of Test 3(a) and (b), determine the input impedance to the transformer.

$Z_{in(measured)}$ =

d. For an ideal transformer, input impedance is $Z_p = a^2\, Z_L$. However, real transformers have winding resistance and leakage reactance. Neglecting leakage reactance, we get the circuit of Figure 32-6. The winding resistance can now be reflected into the primary along with the load resistance. Adding to this the primary winding resistance, we get for the input impedance Z_{in} of the transformer

$$Z_{in} = R_p + a^2(R_s + R_L) \qquad (32\text{-}6)$$

Measurements. Measure primary and secondary winding resistances R_p and R_s using an ohmmeter, then use Equation 32-6 to compute the input impedance for the transformer. Compare to the value determined in (c) i.e., calculate the percent difference. If they differ, what are the likely sources of the difference?

Z_{in} = _____

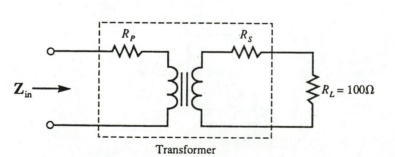

Figure 32-6   Transformer circuit with winding resistance included

PART D: Frequency Response of a Transformer

4. a. Using the audio transformer, reconnect the circuit of Figure 32-3, but add a 100 Ω load resistor between points *c* and *d*. Maintain the input voltage constant at 6 V (i.e., 12 V_{p-p}) and measure the output voltage at the frequencies listed in Table 32-2.

    b. Compute the ratio of output voltage to input voltage over the range of frequencies tested and plot as Figure 32-7. Note that the plot drops off at each end.

f	V_S
100 Hz	
500 Hz	
1 kHz	
5 kHz	
10 kHz	
15 kHz	
20 kHz	

Table 32-2

PART E: Load Voltage Regulation

5. Ideally, a transformer will deliver constant output voltage. In reality, due to internal voltage drops, the output voltage falls as load current increases. In this test, we look at how badly the transformer voltage drops. Here, the transformer input voltage is 120 VAC.

    a. Connect a 120-VAC plug and cord to the power transformer of

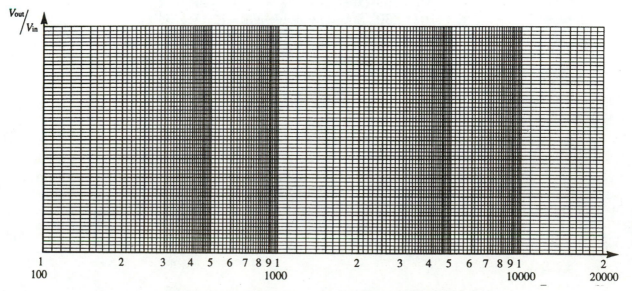

Figure 32-7 Frequency response for an audio transformer

Test 1. (Note: No measurements will be made on the 120-V side. Therefore, make sure that all wiring on the 120-V side is completely insulated for safety. Check with your instructor to be sure.)

b. With a meter, measure the open circuit (no-load) voltage on the 12.6-V winding.

V_{NL} = _____

c. Connect a load resistor that will draw full rated current (or nearly full-rated current). (If you use the Hammond 167L12 as recommended, you will need about 6 Ω, capable of dissipating 25 W. Since this is not a standard resistor, use a 10 Ω resistor instead. You will easily observe the regulation.) Measure output voltage under load. V_{FL} = _____

d. Compute the regulation of the transformer using the formula

$$\text{Regulation} = \left(\frac{V_{NL} - V_{FL}}{V_{FL}} \right) \times 100\%$$

(In general, regulation is undesirable. In later electronic courses, you will learn how to compensate for transformer regulation in power supply circuits.)

PROBLEMS

6. An ideal transformer with a turns ratio of $a = 5$ has a load of $\mathbf{Z}_L = 10\,\Omega\,\angle 30°$. Source voltage is $\mathbf{E} = 120\text{ V } \angle 0°$.
   a. Compute the load voltage and load current as in Example 24-4 of the text.
   b. Using Equation 32-3, compute source current.
   c. Compute the input impedance $\mathbf{Z}_p$.
   d. Using $\mathbf{Z}_p$, compute source current and compare to the result obtained in (b).
7. The ideal transformer has a flat frequency response. However, as Test D indicated, the output for a real transformer falls off at the low end and at the high end of the frequency range of the transformer. Briefly discuss why.

NAME _____

DATE _____

CLASS _____

Mutual Inductance and Loosely Coupled Circuits

OBJECTIVE

After completing this lab, you will be able to
- determine mutual inductance experimentally,
- measure mutual voltage and verify theoretically,
- compare calculated and measured results for coupled circuits.

EQUIPMENT REQUIRED

☐ Oscilloscope, dual channel
☐ Signal or function generator
☐ Inductance meter or bridge

COMPONENTS

☐ 455 kHz IF (*intermediate frequency*) transformer
☐ Resistor: 10-Ω, 1/4-W
☐ Capacitors: 1-µF, non-electrolytic (two required)
Note: For PSpice and Electronics Workbench versions of this lab, see Lab #37.

EQUIPMENT USED

Instrument	Manufacturer/Model No.	Serial No.
Oscilloscope		
Signal or function generator		
Inductance meter or bridge		

Table 33-1

TEXT REFERENCE

DISCUSSION

If two coils with mutual coupling are connected in series so that their fluxes add, their total inductance is

$$L_T{}^+ = L_1 + L_2 + 2\,M \tag{33-1}$$

where L_1 and L_2 are the self-inductances of coils *1* and *2* as indicated in Figure 33-1, and M is the mutual coupling between them. If the coils are connected so that their fluxes subtract as in (b), total inductance is

$$L_{T-} = L_1 + L_2 - 2\,M \tag{33-2}$$

The mutual inductance between the coils can be found from the formula

$$M = (L_T{}^+ - L_T{}^-)/4 \tag{33-3}$$

which is obtained by subtracting Equation 33-2 from 33-1. (Since inductances L_1, L_2, L_{T+} and L_{T-} can be measured with a standard inductance meter, you can use Equation 33-3 to determine mutual inductance experimentally.) Self and mutual inductances are related by the formula

$$M = k\sqrt{L_1 L_2} \tag{33-4}$$

where k is called the *coefficient of coupling*. For tightly coupled coils, $k \approx 1$; for loosely coupled coils, k is much less than 1.

Equations for Coupled Circuits. If coupled coils are connected as in Figure 33-2(a), their voltages and currents are related by the formulas

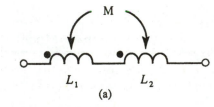

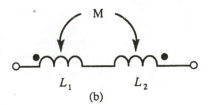

(a) Additive fluxes   *(b) Subtractive fluxes*

Figure 33-1   Coupled coils

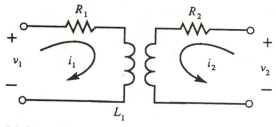 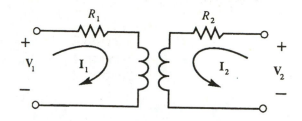

(a) General   (b) Sinusoidal ac

Figure 33-2   Coupled coils

$$v_1 = R_1 i_1 + L_1 \frac{di_1}{dt} \pm M \frac{di_2}{dt} \tag{33-5}$$

$$v_2 = \pm M \frac{di_1}{dt} + R_2 i_2 + L_2 \frac{di_2}{dt} \tag{33-6}$$

where the sign of M is determined by the dot convention—i.e., if both currents enter (or leave) dotted terminals, then the sign to use with M is positive, while if one current enters a dotted terminal and the other leaves, the sign to use is negative. For sinusoidal excitation as shown in Figure 33-2(b), Equations 33-5 and 33-6 become

$$\mathbf{V}_1 = (R_1 + j\omega L_1)\mathbf{I}_1 \pm j\omega M \mathbf{I}_2 \tag{33-7}$$

$$\mathbf{V}_2 = \pm j\omega M \mathbf{I}_1 + (R_2 + j\omega L_2)\mathbf{I}_2 \tag{33-8}$$

MEASUREMENTS

For this lab, use a 455 kHz IF (*intermediate frequency*) transformer with coils wound on a ferrite core. (Note: Some IF transformers come with small capacitors connected. Make sure you disconnect them.)

PART A: Coil Parameters and Mutual Inductance

1. a. Measure the primary and secondary inductances and resistances using an *LRC* meter or impedance bridge.

$L_1 = $ _____ $R_1 = $ _____

$L_2 = $ _____ $R_2 = $ _____

  b. Mark the coil ends for identification (if they are not already numbered), then connect the primary and secondary coils in series as in Figure 33-1. Measure the inductance of the series connection, then reverse the connections and measure again. (The connection with the largest inductance corresponds to L_T^+.)

L_T^+ _____ $L_T^- = $ _____

Use Equation 33-3 to compute M. $M = $ _____

Use Equation 33-4 to compute k. $k =$ _____

Is this transformer loosely coupled or tightly coupled? _____

c. Use the results of Test 1(b) to determine the dotted ends for the coils. (Mark the dotted ends on the coils as you will need this information in later tests.) Describe how you arrived at this conclusion.

_____

_____

PART B: Mutually Induced Voltage

2. Measure the value of the 10-Ω sensing resistor R_S, then assemble the circuit as in Figure 33-3.

$R_S =$ _____

a. Set the source voltage to 4 V_p (8 V_{p-p}) at $f = 1$ kHz. Measure V_S, magnitude and angle (using the oscilloscope) and record below. Compute input current I_1, magnitude, and angle. (You can leave the results in peak volts and amps, rather than convert to rms.)

$V_S =$ _____ $\angle$ $I_1 =$ _____ $\angle$

b. With Probe 1 still measuring e, move Probe 2 to measure secondary voltage v_2. Measure magnitude and angle.

$V_2 =$ _____ $\angle$

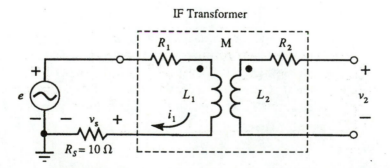

Figure 33-3   Circuit for Test 2. The secondary is open-circuited here.

c. Using the measured values of E, R_1, R_S and L_1, compute current I_1. Thus,

$$I_1 = \frac{E}{R_1 + R_S + j\omega L_1}$$

Compare to the result measured in (a).

d. Compute V_2 from the relationship

$$V_2 = j\omega M I_1 =$$

Compare to the result measured in (b).

PART C: Coupled Circuit Equations

3. Add 2 μF of capacitance to the secondary circuit as in Figure 33-4.
   a. Set e to 4 V_p (8 V_{p-p}). Carefully measure V_S and V_2 as before.

   $V_S =$ _____ ∠ $V_2 =$ _____ ∠

   Compute current $I_1 = V_S/R_1 =$ _____ ∠

   b. Write mesh equations for this circuit and solve for I_1 and I_2. Using the computed value of I_2, calculate V_2. Compare I_1 and V_2 to the values measured in 3(a). How well do they agree?

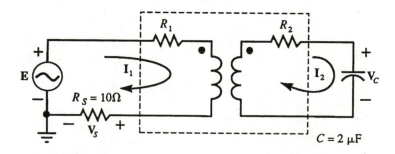

Figure 33-4   Circuit for Test 3

c. Solve for I_1 using the coupled impedance equation

$$Z_{in} = Z_1 + \frac{(\omega M)^2}{Z_2 + Z_L}$$

Compare to the measured value.

PROBLEM

4. Compute I_1, I_2 and V_2 for the circuit of Figure 33-5.

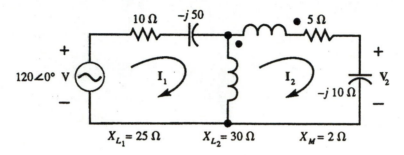

Figure 33-5

LAB 34

Computer Analysis: ac Circuits

OBJECTIVES

After completing this lab, you will be able to
- use PSpice or Electronics Workbench to obtain currents and voltages for an ac circuit,
- use PSpice or Electronics Workbench to find the Thévenin equivalent of an ac circuit,
- compare the predicted results to actual measured values.

EQUIPMENT REQUIRED

☐ Personal computer with appropriate software (OrCAD PSpice or Electronics Workbench)

TEXT REFERENCE

Section 19.7   CIRCUIT ANALYSIS USING COMPUTERS
Section 20.7   CIRCUIT ANALYSIS USING COMPUTERS

DISCUSSION

In Lab 23 you constructed an ac circuit consisting of a sinusoidal voltage source and several components. We begin our analysis by examining how PSpice will solve for the currents and voltages in this simple circuit. Since PSpice provides all voltages and currents as phasor quantities, it is necessary to first convert any voltage sources from time domain into phasor domain. The expressions of (34-1) and (34-2) are equivalent forms of expressing the same sinusoidal waveform; (34-1) is in time-domain while (34-2) is in phasor-domain.

$$v = V_m\sin(\omega t + \theta) \tag{34-1}$$

$$\mathbf{V} = \frac{V_m}{\sqrt{2}} \angle h$$

(34-2)

In Lab 24, you observed how to measure the Thévenin (open circuit) voltage and the Norton (short circuit) current of an ac circuit. The Thévenin (or Norton) impedance of the circuit was determined as the ratio of the Thévenin voltage and the Norton current.

$$\mathbf{Z}_{\mathrm{Th}} = \frac{\mathbf{E}_{\mathrm{Th}}}{\mathbf{I}_{\mathrm{N}}}$$

(34-3)

In this lab, you will use these important principles to determine the Thévenin (and Norton) equivalent circuits of circuits consisting of dependent sources. Unlike other ac circuits, the Thévenin or Norton impedance of a circuit containing dependent sources cannot be found by simply looking into the open terminals.

OrCAD PSpice

When using the OrCAD Capture program to analyze a circuit it is necessary first to use the VPRINT1 part to measure the open-circuit voltage. It is necessary to activate the AC, MAG, and PHASE cells of this part by entering **OK** into each of the cells in the Properties dialog box. Since the circuit consists of only resistors, any suitable frequency can be used to provide the AC Sweep. Since the Probe postprocessor is not required for this lab, it is convenient to disable this feature prior to simulating the design. The following parts in OrCAD Capture are used to represent dependent sources:

Voltage-controlled voltage source: **E**
Current-controlled current source: **F**
Voltage-controlled current source: **G**
Current-controlled voltage source: **H**

You will need to use the IPRINT part to measure the short-circuit current between terminals *a* and *b*.

Electronics Workbench

Electronics Workbench uses a similar approach to simulating circuits having dependent sources. Rather than using VPRINT and IPRINT to obtain the voltage and current, it is a simple matter of using a DMM and selecting either the ac volts range to measure the open circuit voltage or the ac current range to measure the short circuit current. Once again, the circuit frequency is not important. Use equation (34-3) to determine the Thévenin or Norton impedance of a circuit from the measured values.

1. Use either simulation program to create the circuit of Figure 34-1.

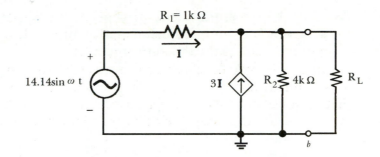

Figure 34-1   Circuit with current-dependent current source

2. Determine the open-circuit voltage between terminals *a* and *b*. Enter the result here.

E_{Th}	

3. Determine the short-circuit voltage between terminals *a* and *b*. Enter the result here.

I_N	

4. Calculate the Thévenin or Norton impedance and enter the value here.

I_N	

5. Sketch the Thévenin and Norton equivalent circuits in this space:

6. Use the appropriate circuit theory to calculate the Thévenin equivalent of the circuit shown in Figure 34-1. Compare the computer results to the theoretical results.

NAME _____

DATE _____

CLASS _____

LAB 35

Computer Analysis: *RC* and *RL* Transients

OBJECTIVE

After completing this lab, you will be able to
- use PSpice or Electronics Workbench to solve simple *RC* circuits,
- use PSpice or Electronics Workbench to solve simple *RL* circuits.

EQUIPMENT REQUIRED

☐ Personal computer with appropriate software (OrCAD PSpice or Electronics Workbench).

TEXT REFERENCE

Section 11.8   TRANSIENT ANALYSIS USING COMPUTERS
Section 14.6   *RL* TRANSIENTS USING COMPUTERS

DISCUSSION

In Lab 13, Part D, you studied capacitor charging and discharging using an oscilloscope. A function generator was used to apply 5 V for 0.5 ms and 0 V for the next 0.5 ms as illustrated in Figure 35-1(a) (next page). The equivalent switching circuit is shown in (b).

We will now simulate this lab using software. Follow the instructions below, using either OrCAD PSpice or Electronics Workbench as applicable. In each case, remember to set the initial condition for the capacitor to zero and to perform a transient analysis.

OrCAD PSpice

Create the circuit of Figure 35-2 (next page) using source VPULSE with parameters V1, V2, PW, PER, TR, TF and TD set respectively to OV, 5V, 0.5ms, 1ms, 1ns, 1ns and 0. (This creates a pulse with an ini-

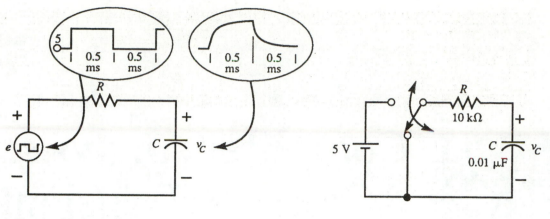

(a) Circuit of Lab 13, Test 5

(b) Equivalent circuit

Figure 35-1 Capacitor charging and discharging

tial value of 0 V, an amplitude of 5 V, a pulse width of 0.5 ms, a period of 1 ms, 1ns rise and fall times and a delay to start of 0 s to simulate the waveform of Fig. 35-1.) Now proceed to "PART A: Simulating an *RC* Transient" to test the circuit.

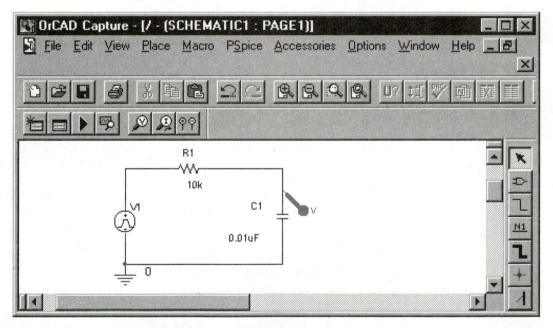

Figure 35-2

Electronics Workbench

Create the circuit of Figure 35-3 using EWB's square wave clock source with *f* set to 1000 Hz, amplitude 5 V. [Note: The clock waveform is low during the first half cycle and high during the second half (which is opposite to what we want). The circuit of Figure 35-3 provides the desired waveform.]

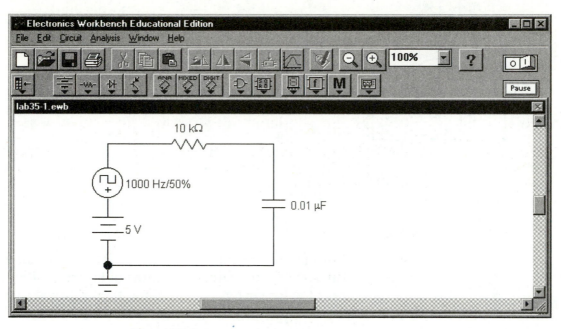

Figure 35-3

PART A: Simulating an *RC* Transient

1. a. Create the circuit of Figure 35-2 or 35-3 as applicable. Run the simulation to 1 ms (to record both charging and discharging) and obtain a plot of capacitor voltage versus time. Print both the circuit schematic and the waveform on your printer and attach to this lab for submission.

   b. Compute the time constant τ for the circuit: $\tau =$ _____.

   c. Using the cursor, read values from the screen at $t = \tau, 2\tau, 3\tau, 4\tau$ and 5τ and tabulate.

   d. In Lab 13, you measured capacitor voltage at these same values of time. Compare the values measured using the scope to those determined in c). How well do they agree?

   e. Repeat Parts c) and d) at $t = \tau, 2\tau, 3\tau, 4\tau$ and 5τ into discharge.

PART B: Simulating an *RL* Transient

2. Consider Figure 35-4(a). (This is the circuit you used in Lab 14, Figure 14-6.) The function generator applies source voltage E to the circuit for 12.5 μs, then 0 V for 12.5 μs, resulting in the current waveform shown in (b).

   To create this waveform, you must determine the value of source voltage needed to create a current of 5 mA. First, obtain the value of coil resistance R_ℓ and signal generator resistance R_{out} from Lab 14. Now compute the required pulse amplitude. For example, if $R_\ell = 12\ \Omega$, $R_{out} = 50\ \Omega$ and $R_s = 1000\ \Omega$, then the source voltage needed is (5 mA) (1062 Ω) = 5.31 V. Now select PSpice or Electronics Workbench as applicable.

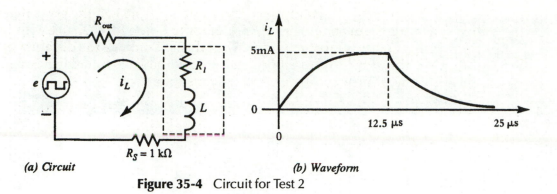

(a) *Circuit*   (b) *Waveform*

Figure 35-4   Circuit for Test 2

OrCAD PSpice

Use VPULSE as in Part A, except set V2 to 5.31V (or as computed above) and PW and PER to match the timing of Figure 35-4(b). Use a current marker to plot current versus time, then proceed to Procedure step a) below.

Electronics Workbench

Use the same source as in Part A, except set the source amplitude to 5.31 V (or as computed above) and select a frequency to match the timing requirements of Figure 35-3(b). Since EWB cannot easily display current, use the procedure that you used in Lab 14, i.e., plot the voltage across the sense resistor, then compute current using Ohm's law. (Since $R_s = 1$ kΩ, the plot can be interpreted directly in mA, i.e., 5 V represents 5 mA, etc.)

Procedure

a. Build the circuit on your screen using PSpice or EWB as applicable. Set the initial condition (i.e., initial inductor current) to zero and run a transient analysis. (Make sure you get the complete build-up/decay waveform.) Print both the circuit schematic and the waveform on your printer and attach to this lab for submission.
b. Compute the circuit's time constant τ, then position the cursor at $t = \tau$ on the screen and read the current. Compare its value to that measured in Lab 14. Discuss any likely sources of difference.
c. Repeat Part b) at 1 time constant into decay.

PROBLEMS

3. a. Create the circuit of Figure 13-8 (Lab 13) on your screen and plot a curve of capacitor voltage versus time. Print the waveform and attach to this lab for submission.
   b. Reduce the circuit of Figure 13-8 to its Thévenin equivalent. Create the equivalent circuit on your screen and repeat Part a). How well do the two waveforms agree?

4. A theoretical analysis of Figure 14-8(c) from Lab 14 shows that the transient should last 9.3 s. Compute L_1 and R_2, then use your computer to perform a transient analysis. Compare the duration of the transient to the theoretical value. Show all calculations and attach the waveform to the lab for submission.

NAME _____

DATE _____

CLASS _____

LAB 36

Computer Analysis: Resonance

OBJECTIVES

After completing this lab, you will be able to
- use PSpice or Electronics Workbench to provide frequency responses of resonant circuits,
- use cursors to determine the 3-dB down frequencies of resonant circuits,
- compare the predicted results to actual measured values.

EQUIPMENT REQUIRED

☐ Personal computer with appropriate software (OrCAD PSpice or Electronics Workbench)

TEXT REFERENCE

Section 21.7   CIRCUIT ANALYSIS USING COMPUTERS

DISCUSSION

In Lab 25 you measured the frequency response of a series resonant circuit. In order to get an accurate response curve, it was necessary to adjust the frequency of the signal generator and take numerous measurements. Both PSpice and Electronics Workbench can provide a graphical display of the circuit response characteristics as a function of frequency. By using cursors, we are able to determine the maximum output voltage (at resonance) and find the half-power frequencies. Recall that the half-power (or 3-dB down) frequencies occur when the output voltage is 0.707 of the maximum resonant voltage. The instructions that follow will help you simulate the operation of the resonant circuits.

Electronics Workbench

You will need to set the amplitude of the signal generator to its rms value, namely 1.414 V. Since the coil resistance will vary from one inductor to the next, you may use the measured value from either Lab 25 or Lab 26. Alternatively a nominal value of 5 Ω may be used. Select the Bode Plotter from the Instruments tool bin. Connect the input of the plotter to the voltage source and the output across the inductor (including the coil resistance R_{coil}). Run the simulation and double click on the Bode Plotter symbol. You will see a frequency response over a very wide frequency range. You may select a narrower range by choosing initial (I) and final (F) frequencies closer to the resonant frequency. After once again running the simulation, you may click on the Display Graphs tool to view a larger display of the frequency response.

OrCAD PSpice

Use OrCAD Capture CIS to create the circuit. Use the VAC voltage source and set the magnitude for 1.414V (no spaces). Set the simulation settings for an ac sweep from 1kHz to 10 kHz with a total of 10001 points per decade. Click on the Run tool, and the PROBE postprocessor will appear after the calculations are complete. Select the cursor from the Trace menu item.

PART A: Series Resonant Circuit

1. Use either simulation program to create the series resonant circuit shown in Figure 36-1.
2. Use the cursors to determine the resonant frequency and the half-power frequencies. Calculate the bandwidth and the quality factor of the circuit. Enter the values in the spaces provided (next page):

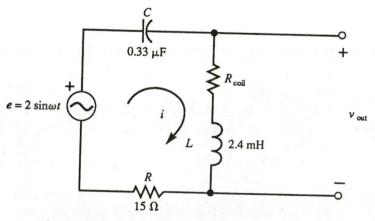

Figure 36-1   Series resonance circuit

f_s	
f_1	
f_2	
BW	
Q	

3. Compare the above values to the measurements of Lab 25. How well do they agree?

_____

_____

4. If a printer is available, obtain a printout of the frequency response, showing voltage as a function of frequency.

PART B: Parallel Resonant Circuit

5. Use either simulation program to create the series resonant circuit shown in Figure 36-2.
6. Use the cursors to determine the resonant frequency and the half-power frequencies. Calculate the bandwidth and the quality factor of the circuit. Enter the values in the space provided (next page):

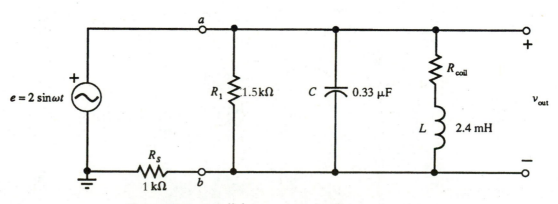

Figure 36-2 Parallel resonance circuit

f_p	
f_1	
f_2	
BW	
Q	

7. Compare the above values to the measurements of Lab 26. How well do they agree?

_____

_____

8. If a printer is available, obtain a printout of the frequency response, showing voltage as a function of frequency.

LAB 37

Computer Analysis: Coupled Circuits

OBJECTIVE

After completing this lab, you will be able to
- use PSpice or Electronics Workbench to analyze coupled circuits.

EQUIPMENT REQUIRED

☐ Personal computer with appropriate software (OrCAD PSpice or Electronics Workbench)

TEXT REFERENCE

Section 24.12 CIRCUIT ANALYSIS USING COMPUTERS

DISCUSSION

We now look at the analysis of coupled circuits using PSpice and Electronics Workbench. (PSpice can handle both loosely coupled and tightly coupled circuits, but, as of this writing, EWB has no simple way to handle loosely coupled circuits. We therefore limit our attention to iron-core circuits for Electronics Workbench.) In what follows, first read the appropriate section (PSpice or Electronics Workbench), then proceed to the Lab simulation section.

OrCAD PSpice

As you saw in Chapter 24, loosely coupled circuits are modeled by self and mutual inductances L_1, L_2 and M (or L_1, L_2 and k where $k = \frac{M}{\sqrt{L_1 L_2}}$), while tightly coupled circuits (such as the iron-core transformer), are modeled by their turns ratio a. With SPice, the first case is straightforward, as PSpice uses L_1, L_2 and k. The second case, however, requires a bit of thought. First, note that for an ideal, tightly coupled transformer, L_1 and L_2 are infinite, but their ratio is finite and equal to $L_1/L_2 = a^2$. (This may be shown theoretically.) Thus, to model an iron-core transformer, set $k = 1$, arbitrarily choose a very large value for L_1, then compute L_2 as noted. For example, if you have a transformer with a turns ratio of 5, set L_1 to an arbitrarily large value (e.g., 25 000 H), then compute $L_2 = L_1/a^2 = 25\,000/(5^2) = 1000$ H. (The value chosen for L_1 is not critical, it simply must be very large.) Now set $k = 1$ and PSpice treats this as an ideal transformer with $a = 5$.

The PSpice circuit element that we are going to use here is designated XFRM_LINEAR. XFRM_LINEAR handles phasing by means of the dot convention, Figure 37-1. Note that the sign of k carries the phasing information. For example, if the coupling coefficient is 0.5 and both dots are at the same end of the winding as in (a), set $k = 0.5$, while if they are at opposite ends as in (b), set $k = -0.5$.

Electronics Workbench

The circuit element that we will use for EWB is shown in Figure 37-2. (This device models transformers based on their turns ratio. However, as noted in Section 24-12 of the textbook, it also includes a parameter L_m that you must set. To approximate an ideal transformer, L_m must be very large. A value of 10 000 H will serve for this lab.)

PART A: Regulation of an Iron-Core Transformer

1. An iron-core transformer with primary and secondary resistance and leakage reactance is shown in Figure 37-3. (R_p is the resistance of the primary coil winding and R_s is the resistance of the secondary. Inductances L_p and L_s represent the primary and secondary leakage flux respectively.) The transformer has a 10:1 turns ratio and is driven from a 120-V source. At no load, the voltage will be 12 V, but for the load shown, V_{load} will be somewhat less. Frequency is 60 Hz.

   a. Using PSpice or EWB as applicable, simulate the circuit and de-

   termine the load voltage at full load. $V_{load(FL)} =$ _____

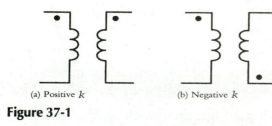

(a) Positive k (b) Negative k

Figure 37-1

Figure 37-2

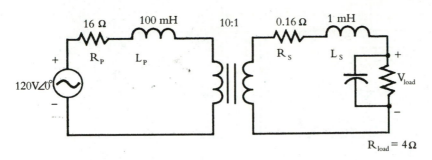

Figure 37-3

    b. Repeat the simulation at no load. (Note: You may have to re-place the load with a large value resistance, e.g., 100 kΩ to sim-

        ulate the open circuit.) $V_{load(NL)}$ = _____

    c. Using the values determined in b) and c), compute regulation.
    d. Solve the problem manually (i.e., with your calculator) using the techniques of Chapter 24, Section 24.6, and compare to the answers of a), b) and c).

PART B: Voltages and Currents in a Loosely Coupled Circuit

(PSpice users only),

  2. Consider Figure 37-4. (This is the circuit that you used in Lab 33, Figure 33-3.) Using parameter values from Lab 33, simulate this problem using PSpice and determine V_2. Compare to the voltage measured in Lab 33.
  3. Consider Figure 37-5. (This is the circuit that you used in Lab 33, Figure 33-4.) Using parameter values from Lab 33, simulate this problem using PSpice and determine current I_2 and voltage V_2. Compare to the values measured in Lab 33.

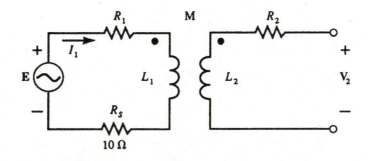

Figure 37-4 Circuit for Test 2

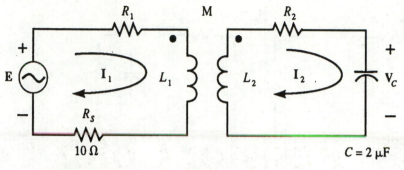

Figure 37-5   Circuit for Test 3

PROBLEM

4. Use PSpice and solve for $\mathbf{I}_1$, $\mathbf{I}_2$ and $\mathbf{V}_2$ of Figure 37-6. Given $L_1 = 5$ mH, $L_2 = 6$ mH, $M = 0.4$ mH, and $f = 1000$ Hz.

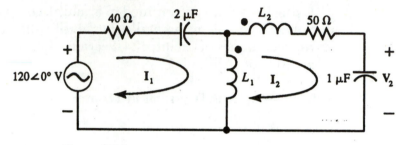

Figure 37-6

Resistor Color Codes

Figure A-1 shows a typical color coded resistor together with the standard colors and corresponding designations. Table A-1 gives nominal resistor values for resistors having 5%, 10%, and 20% values.

First Band — 1st Digit

Second Band — 2nd Digit

Color	Digit	Multiplier	Tolerance	Reliability Level (Percent Per 1000 Hours) (When Applicable)
Black	0	1	—	—
Brown	1	10	—	M = 1.0%
Red	2	100	—	P = 0.1%
Orange	3	1000	—	R = 0.01%
Yellow	4	10,000	—	S = 0.001%
Green	5	100,000	—	—
Blue	6	1,000,000	—	—
Violet	7	10,000,000	—	—
Gray	8	—	—	—
White	9	—	—	—
Gold	—	0.1	± 5%	—
Silver	—	—	± 10%	—
No color	—	—	± 20%	—

Figure A-1 Standard resistor values (Courtesy of the Allen-Bradley Company)

Resistance Color Code — Nominal Resistance in Ohms

1st BAND (1st digit)	2nd BAND (2nd digit)	3rd BAND (Number of zeros after 1st and 2nd digit)	Gold ±5%	Silver ±10%	None ±20%
Brown	Black	Gold	1.0	1.0	1.0
Brown	Brown	Gold	1.1	—	—
Brown	Red	Gold	1.2	1.2	—
Brown	Orange	Gold	1.3	—	—
Brown	Green	Gold	1.5	1.5	1.5
Brown	Blue	Gold	1.6	—	—
Brown	Gray	Gold	1.8	1.8	—
Red	Black	Gold	2.0	—	—
Red	Red	Gold	2.2	2.2	2.2
Red	Yellow	Gold	2.4	—	—
Red	Violet	Gold	2.7	2.7	—
Orange	Black	Gold	3.0	—	—
Orange	Orange	Gold	3.3	3.3	3.3
Orange	Blue	Gold	3.6	—	—
Orange	White	Gold	3.9	3.9	—
Yellow	Orange	Gold	4.3	—	—
Yellow	Violet	Gold	4.7	4.7	4.7
Green	Brown	Gold	5.1	—	—
Green	Blue	Gold	5.6	5.6	—
Blue	Red	Gold	6.2	—	—
Blue	Gray	Gold	6.8	6.8	6.8
Violet	Green	Gold	7.5	—	—
Gray	Red	Gold	8.2	8.2	—
White	Brown	Gold	9.1	—	—
Brown	Black	Black	10	10	10
Brown	Brown	Black	11	—	—
Brown	Red	Black	12	12	—
Brown	Orange	Black	13	—	—
Brown	Green	Black	15	15	15
Brown	Blue	Black	16	—	—
Brown	Gray	Black	18	18	—
Red	Black	Black	20	—	—
Red	Red	Black	22	22	22
Red	Yellow	Black	24	—	—
Red	Violet	Black	27	27	—
Orange	Black	Black	30	—	—
Orange	Orange	Black	33	33	33
Orange	Blue	Black	36	—	—
Orange	White	Black	39	39	—
Yellow	Orange	Black	43	—	—
Yellow	Violet	Black	47	47	47
Green	Brown	Black	51	—	—
Green	Blue	Black	56	56	—
Blue	Red	Black	62	—	—
Blue	Gray	Black	68	68	68
Violet	Green	Black	75	—	—
Gray	Red	Black	82	82	—
White	Brown	Black	91	—	—
Brown	Black	Brown	100	100	100
Brown	Brown	Brown	110	—	—
Brown	Red	Brown	120	120	—
Brown	Orange	Brown	130	—	—
Brown	Green	Brown	150	150	150
Brown	Blue	Brown	160	—	—
Brown	Gray	Brown	180	180	—
Red	Black	Brown	200	—	—
Red	Red	Brown	220	220	220
Red	Yellow	Brown	240	—	—
Red	Violet	Brown	270	270	—
Orange	Black	Brown	300	—	—
Orange	Orange	Brown	330	330	330
Orange	Blue	Brown	360	—	—
Orange	White	Brown	390	390	—
Yellow	Orange	Brown	430	—	—
Yellow	Violet	Brown	470	470	470

Resistance Color Code — Nominal Resistance in Ohms

1st BAND (1st digit)	2nd BAND (2nd digit)	3rd BAND (Number of zeros after 1st and 2nd digit)	Gold ±5%	Silver ±10%	None ±20%
Green	Brown	Brown	510	—	—
Green	Blue	Brown	560	560	—
Blue	Red	Brown	620	—	—
Blue	Gray	Brown	680	680	680
Violet	Green	Brown	750	—	—
Gray	Red	Brown	820	820	—
White	Brown	Brown	910	—	—
Brown	Black	Red	1000	1000	1000
Brown	Brown	Red	1100	—	—
Brown	Red	Red	1200	1200	—
Brown	Orange	Red	1300	—	—
Brown	Green	Red	1500	1500	1500
Brown	Blue	Red	1600	—	—
Brown	Gray	Red	1800	1800	—
Red	Black	Red	2000	—	—
Red	Red	Red	2200	2200	2200
Red	Yellow	Red	2400	—	—
Red	Violet	Red	2700	2700	—
Orange	Black	Red	3000	—	—
Orange	Orange	Red	3300	3300	3300
Orange	Blue	Red	3600	—	—
Orange	White	Red	3900	3900	—
Yellow	Orange	Red	4300	—	—
Yellow	Violet	Red	4700	4700	4700
Green	Brown	Red	5100	—	—
Green	Blue	Red	5600	5600	—
Blue	Red	Red	6200	—	—
Blue	Gray	Red	6800	6800	6800
Violet	Green	Red	7500	—	—
Gray	Red	Red	8200	8200	—
White	Brown	Red	9100	—	—
Brown	Black	Orange	10000	10000	10000
Brown	Brown	Orange	11000	—	—
Brown	Red	Orange	12000	12000	—
Brown	Orange	Orange	13000	—	—
Brown	Green	Orange	15000	15000	15000
Brown	Blue	Orange	16000	—	—
Brown	Gray	Orange	18000	18000	—
Red	Black	Orange	20000	—	—
Red	Red	Orange	22000	22000	22000
Red	Yellow	Orange	24000	—	—
Red	Violet	Orange	27000	27000	—
Orange	Black	Orange	30000	—	—
Orange	Orange	Orange	33000	33000	33000
Orange	Blue	Orange	36000	—	—
Orange	White	Orange	39000	39000	—
Yellow	Orange	Orange	43000	—	—
Yellow	Violet	Orange	47000	47000	47000
Green	Brown	Orange	51000	—	—
Green	Blue	Orange	56000	56000	—
Blue	Red	Orange	62000	—	—
Blue	Gray	Orange	68000	68000	68000
Violet	Green	Orange	75000	—	—
Gray	Red	Orange	82000	82000	—
White	Brown	Orange	91000	—	—

Nominal Resistance in Megohms

1st digit	2nd digit	zeros	Gold ±5%	Silver ±10%	None ±20%
Brown	Black	Yellow	0.1	0.1	0.1
Brown	Brown	Yellow	0.11	—	—
Brown	Red	Yellow	0.12	0.12	—
Brown	Orange	Yellow	0.13	—	—
Brown	Green	Yellow	0.15	0.15	0.15
Brown	Blue	Yellow	0.16	—	—
Brown	Gray	Yellow	0.18	0.18	—
Red	Black	Yellow	0.20	—	—
Red	Red	Yellow	0.22	0.22	0.22
Red	Yellow	Yellow	0.24	—	—

Resistance Color Code — Nominal Resistance in Megohms

1st BAND (1st digit)	2nd BAND (2nd digit)	3rd BAND (Number of zeros after 1st and 2nd digit)	Gold ±5%	Silver ±10%	None ±20%
Red	Violet	Yellow	0.27	0.27	—
Orange	Black	Yellow	0.30	—	—
Orange	Orange	Yellow	0.33	0.33	0.33
Orange	Blue	Yellow	0.36	—	—
Orange	White	Yellow	0.39	0.39	—
Yellow	Orange	Yellow	0.43	—	—
Yellow	Violet	Yellow	0.47	0.47	0.47
Green	Brown	Yellow	0.51	—	—
Green	Blue	Yellow	0.56	0.56	—
Blue	Red	Yellow	0.62	—	—
Blue	Gray	Yellow	0.68	0.68	0.68
Violet	Green	Yellow	0.75	—	—
Gray	Red	Yellow	0.82	0.82	—
White	Brown	Yellow	0.91	—	—
Brown	Black	Green	1.0	1.0	1.0
Brown	Brown	Green	1.1	—	—
Brown	Red	Green	1.2	1.2	—
Brown	Orange	Green	1.3	—	—
Brown	Green	Green	1.5	1.5	1.5
Brown	Blue	Green	1.6	—	—
Brown	Gray	Green	1.8	1.8	—
Red	Black	Green	2.0	—	—
Red	Red	Green	2.2	2.2	2.2
Red	Yellow	Green	2.4	—	—
Red	Violet	Green	2.7	2.7	—
Orange	Black	Green	3.0	—	—
Orange	Orange	Green	3.3	3.3	3.3
Orange	Blue	Green	3.6	—	—
Orange	White	Green	3.9	3.9	—
Yellow	Orange	Green	4.3	—	—
Yellow	Violet	Green	4.7	4.7	4.7
Green	Brown	Green	5.1	—	—
Green	Blue	Green	5.6	5.6	—
Blue	Red	Green	6.2	—	—
Blue	Gray	Green	6.8	6.8	6.8
Violet	Green	Green	7.5	—	—
Gray	Red	Green	8.2	8.2	—
White	Brown	Green	9.1	—	—
Brown	Black	Blue	10	10	10
Brown	Brown	Blue	11	—	—
Brown	Red	Blue	12	12	—
Brown	Orange	Blue	13	—	—
Brown	Green	Blue	15	15	15
Brown	Blue	Blue	16	—	—
Brown	Gray	Blue	18	18	—
Red	Black	Blue	20	—	—
Red	Red	Blue	22	22	22
Red	Yellow	Blue	24	—	—
Red	Violet	Blue	27	27	—
Orange	Black	Blue	30	—	—
Orange	Orange	Blue	33	33	33
Orange	Blue	Blue	36	—	—
Orange	White	Blue	39	39	—
Yellow	Orange	Blue	43	—	—
Yellow	Violet	Blue	47	47	47
Green	Brown	Blue	51	—	—
Green	Blue	Blue	56	56	—
Blue	Red	Blue	62	—	—
Blue	Gray	Blue	68	68	68
Violet	Green	Blue	75	—	—
Gray	Red	Blue	82	82	—
White	Brown	Blue	91	—	—
Brown	Black	Violet	100	100	100

Table A-1 Nominal resistor values (Courtesy of the Allen-Bradley Company)

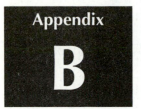

Equipment and Components

Standard Equipment for Basic Labs

1—Dual channel oscilloscope
1—DMM (Occasionally two are required)
1—VOM (Occasionally two are required)
1—Power supply, variable, regulated
1—Signal generator (A function generator is required for a few labs.)
1—*LRC* meter or impedance bridge (for Labs 12 and 14 only)

Specialized Equipment for the Power Labs

2—Single-phase wattmeters (approximately 300 W full scale)
Three-phase ac source, 120/208 V

Resistors

Most resistor tolerances are not critical. However, unless otherwise noted, we recommend 5% tolerance resistors or better.

Resistors (1/4 W)

$4.7\,\Omega$, $6.8\,\Omega$, $10\,\Omega$, $15\,\Omega$, $47\,\Omega$, $75\,\Omega$, $82\,\Omega$, $100\,\Omega$, $150\,\Omega$, $220\,\Omega$, $270\,\Omega$, $330\,\Omega$, $470\,\Omega$, $680\,\Omega$, $1\,k\Omega$, $1.5\,k\Omega$, $2\,k\Omega$, $2.2\,k\Omega$, $2.7\,k\Omega$, $3.3\,k\Omega$, $3.9\,k\Omega$, $4.7\,k\Omega$, $5.6\,k\Omega$, $6.8\,k\Omega$, $10\,k\Omega$, $20\,k\Omega$, $39\,k\Omega$, $47\,k\Omega$, $100\,k\Omega$, $180\,k\Omega$, $330\,k\Omega$, $3.3\,M\Omega$, $5.6\,M\Omega$

Resistors (Other)

(1/8 W)	470 Ω
(1/2 W)	47 Ω, 82 Ω, 470 Ω
(1 W)	470 Ω
(2 W)	47 Ω, 75 Ω, 82 Ω, 100 Ω, 120 Ω, 270 Ω, 470 Ω
(25 W)	10 Ω (Lab 32)
(200 W)	100 Ω (One for Lab 20. If you do the three phase labs, you will need three.)
(200-W)	250 Ω (Three. Required for the three phase labs only.)

Capacitors

2200 pF, 3300 pF, 0.01 μF, 0.047 μF, 0.22 μF, 0.33 μF, 0.47 μF, 1.0 μF, all ± 10%

Capacitors (Other)

470 μF, electrolytic (For Lab 13. 35 WVDC)
30 μF, non-electrolytic, rated for 120 VAC operation. (One for Lab 20. If you do the three phase labs, you will need three.)

Inductors

1 mH. Powdered iron-core inductor (Hammond #1534A or equivalent)
2.4 mH (two). Powdered iron-core inductor (Hammond #1534C or equivalent. You need an inductor with low resistance to approximate the behavior of an ideal inductor.)
0.2 H (approximate value) inductor. Rated to handle 2 amps.
1.5 H iron-core inductor. (Approximate value. Value not critical.)

Miscellaneous

Batteries: 1.5 V (D cell), 9 V
Thermistor, 10 kΩ @ 25°C
Signal diode: 1N4004 or equivalent
Zener diode: 1N4734A (1 W, 5.6 V, ± 5%)
10 kΩ variable resistor
5 kΩ variable resistor
455 kHz IF transformer
Transformer. 120/12.6 V filament transformer